technology, culture, and development

The Experience of the Soviet Model

technology, culture, and development

The Experience of the Soviet Model

LOREN R. GRAHAM

PAUL R. JOSEPHSON

SUSAN J. LINZ

STEVEN W. POPPER

JOHN M. KRAMER

JOAN DeBARDELEBEN

BARBARA JANCAR-WEBSTER

Edited by JAMES P. SCANLAN

Routledge
Taylor & Francis Group

LONDON AND NEW YORK

First published 1992 by M.E. Sharpe

Published 2015 by Routledge
2 Park Square, Milton Park, Abingdon, Oxon OX14 4RN
711 Third Avenue, New York, NY 10017, USA

Routledge is an imprint of the Taylor & Francis Group, an informa business

Library of Congress Cataloging-in-Publication Data
Technology, culture, and development: the experience of the Soviet model /
edited by James P. Scanlan.
p. cm.
Papers presented at a conference held at Ohio State University, Columbus,
on October 27–27, 1990
ISBN 0-87332-891-4.—ISBN 0-87332-892-2 (pbk.)
1. Technology and state—Soviet Union—Congresses.
2. Technology—Soviet Union—History—Congresses.
I. Scanlan, James P. (James Patrick), 1927–.
T26.S65T43   1992
338.94706–dc20
91-22611
CIP

ISBN 13: 9780873328920 (pbk)
ISBN 13: 9780873328913 (hbk)

Contents

Acknowledgments

The papers in this volume were originally presented at a conference entitled "Technology, Culture, and Development: The Experience of the Soviet Model," held at the Ohio State University, Columbus, Ohio, on October 26 and 27, 1990. The conference, cosponsored by the University Learning Guild and the Center for Slavic and East European Studies at Ohio State, was funded by a grant from the Battelle Endowment for Technology and Human Affairs. The editor and authors wish to express their thanks to the Battelle Endowment for its generous support of this venture.

Introduction

Although scholars have devoted much attention over the years to the impact of technology on society, they have tended to slight the opposite effect—the impact of social influences (broadly understood) on technology. The authors of the essays in the present volume seek to redress this imbalance: in their examination of the "Soviet model" of technological development as exemplified in the former USSR and the former satellite countries of Eastern Europe, they analyze not simply the consequences of technological change but the ways in which social and cultural factors shape the progress and application of technology.

Some observers had assumed that technology and the science on which it depends can flourish as well in closed, undemocratic societies (such as the "Eastern bloc" countries before 1989) as in open, democratic societies—perhaps can even flourish better, as "Sputnik" seemed to demonstrate in the fifties. In the post–Sputnik era it became clear to everyone, however, that science and technology in the Eastern bloc were declining in relation to the West, and there is general agreement among both Eastern and Western specialists that the cause is to be found in the social, economic, political, and cultural environment within which science and technology have existed in Eastern Europe and the USSR. Indeed it was in part to improve that environment and to bring the USSR back onto the world stage as a major player in science and technology that the Soviet Union after 1985 undertook the ambitious program of social change we know as "perestroika."

On that basis alone, the experience of the Soviet model of technological development would serve as a revealing case study of the bearing of social and cultural factors on the fortunes of technology. But of course the dramatic events that have ensued, not only in the USSR but throughout Eastern Europe, have made such a study all the more timely, for they have turned the region into a veritable laboratory of social change. Both the altered relationships of the countries with each other and the radical transformations taking place within each country have created a wealth of new opportunities for examining the complex interplay of social forces in that region, including, preeminently, the effect of political, economic, and cultural factors on technology and science.

The chapter by Loren R. Graham, which begins this volume, provides a needed historical perspective on the present situation in Soviet technology. Graham identifies a pattern of uneven development that has characterized both tsarist and Communist Russia—a pattern he describes as "momentary achievement followed by obsolescence." Russian science and technology, instead of systematically lagging behind the West and coming close to the latter only in the twentieth century, was marked by world-class successes as early as the seventeenth century; but each time the successes were fleeting, since an advanced position once achieved could not be maintained. Both the achievements and the subsequent obsolescence have had their social causes, in Graham's analysis, during the Soviet period as well as before. Neither tsarist nor Soviet Russia proved capable of creating a culture and an economy that could sustain high levels of science and technology.

Paul R. Josephson brings us to the present day with a detailed analysis of the implications of perestroika for technological development in the USSR, with special attention to the ways in which current political and economic reforms are affecting Soviet attitudes toward technology. Through extensive use of interviews and the daily Soviet press as well as technical literature, Josephson catalogs, on the one hand, the changing relations among scientists, their professional organizations, and Party and government institutions, and on the other hand, the changing valuations of various technologies among Soviet scientists and engineers. He argues that, although Soviet professionals no longer have boundless faith in the benefits of "big science," they continue to see science and technology as panaceas: nuclear and space research, formerly regarded as the keys to correcting the Soviet Union's ills, have been replaced by an attachment to other technologies, such as marketization and the U.S. systems of production and scientific organization.

Continuing the focus of the first two chapters on technological development in the Soviet Union, Susan J. Linz examines the economic environment for innovation and how it is likely to change as a consequence of perestroika. In addition to describing the current innovation environment (with the help of interviews with recent émigrés as well as Soviet and Western publications), she evaluates the competing economic plans of Mikhail Gorbachev and his more radical critics, including the "500-day plan" devised by Stanislav Shatalin. Arguing that some form of market-oriented economy is the likely consequence of the reforms, Linz provides an analysis of different managerial innovation strategies and evaluates alternative courses of action designed to foster the development and diffusion of new technology within the Soviet situation. To the extent that perestroika succeeds, she concludes, managers will be confronted with a radically changed economic environment—but one in which cultural factors are unlikely to impede them in their pursuit of innovation.

In the next two chapters, attention shifts to the countries of Eastern Europe. Steven W. Popper employs the metaphor of "after the flood" to characterize the situation of East European scientists and technology practitioners in the post–1989 world, in which "the economic, political, institutional, and cultural foundations upon which their working environment was predicated have been all but swept away." Popper argues that they differ from their Soviet counterparts in their inability to sustain an R&D system independently, and that, consequently, they must become part of the international mainstream of science and technology in order to preserve their own scientific traditions and to operate effectively. He goes on to detail the kinds of social and cultural changes that are required throughout Eastern Europe if those countries are to rejoin the world scientific community, as well as the problems faced in attempting those changes. Much depends, Popper concludes, on the willingness of the East Europeans to accept Western scientific and commercial involvement in their R&D efforts.

A particular problem for Eastern Europe, created by the Soviet model of technological development and now brought to a head by political and economic changes in the region, is the Soviet–East European energy relationship, examined closely in the essay by John M. Kramer. Encouraged by Soviet policy to promote extensive growth through heavy use of energy resources from the USSR, the energy-poor countries of Eastern Europe became increasingly dependent on imports of petroleum, natural gas, and hard coal from the Soviet Union—to the extent that by 1985 all East European states except Romania received more than 75 percent of their total energy imports from that country. This situation was tolerable as long as the USSR provided ample supplies at prices far below the world market level. But when in 1990 the Soviet Union announced its intention of ceasing to subsidize these transactions, the East European countries were faced with an "energy shock" of massive proportions. Kramer examines the origins and the present and future status of the East European energy predicament, detailing the extent of the problem and its implications for technological development. He argues that the cutting off of cheap Soviet energy may in the long run benefit Eastern Europe by impelling it toward technological modernization and closer relations with the West.

In the final two chapters, attention turns to the environment as a critical junction point at which the concerns of technology, culture, and development intersect in the Soviet Union and Eastern Europe. Joan DeBardeleben focuses on ecology and technology in the USSR, describing the ways in which both Marxist ideology and Soviet industrial policy led to technological practices that degraded the quality of life through their impact on the natural environment. She follows the subsequent growth of the environmental movement in the Soviet Union and the ways in which ecological considerations have begun to have an impact on public policy and the direction of technological change, especially after the nuclear accident at Chernobyl in 1986. In two areas in particular—waste reduction in industry and the agricultural use of chemical pesticides and fertilizers—she documents the expression of public pressures for technological conversion. Despite such pressures, however, and despite the possible ecological benefits of the economic reforms now underway in the USSR, DeBardeleben finds that ecological considerations have not yet had a substantial impact on technological development in the country: economic constraints combined with the vast scope of the problem impede the implementation of desired solutions.

In her treatment of technology and the environment in Eastern Europe, Barbara Jancar-Webster begins by sketching a general theoretical framework for considering the relationship between the two and, in particular, the relationship between linear technological development and the nonlinear, unexpected occurrence of environmental disbenefits. Applying this framework to the countries of Eastern Europe, she catalogs the advantages and disadvantages likely to accrue, from an ecological standpoint, from the transition to democratic, market-oriented societies. Although the disadvantages are formidable, she believes they are offset by the advantages, among which is the strong link between democratization and the environmental movement. In a country-by-country survey of the state of that movement throughout the region, Jancar-Webster argues that the movement is in a good position to influence favorably the course of both democracy and technology in all the countries of Eastern Europe. She concludes with the hope that, if the East European governments will ally themselves less with industry and more with consumer and environmental groups, the region may be able to attain sustainable development in a clean environment.

Whatever in fact the future holds for the Soviet Union and Eastern Europe, the social and cultural influences on technological development that are examined here will surely figure in the outcome. Exactly what role these or any influences will play, however, will be determined by the peoples of Eastern Europe and the USSR, who are at this moment coping practically with the difficult social problems the authors have sought to analyze in these essays.

JAMES P. SCANLAN

Contributors

Joan DeBardeleben is Associate Professor at the Institute of Soviet and East European Studies at Carleton University in Ottawa, Canada. She has published widely on environmental questions in the Soviet Union and Eastern Europe. Her publications include articles in scholarly journals and the book *The Environment and Marxism-Leninism: The Soviet and East German Experience* (1985). She is also the author of *Soviet Politics in Transition*, forthcoming in 1992).

Loren R. Graham is Professor of the History of Science at Massachusetts Institute of Technology. He also teaches at Harvard University. A specialist in the history of Russian and Soviet science and technology, he has published many books and articles on those subjects. His recent book *Science, Philosophy, and Human Behavior in the Soviet Union* (1987) was published in 1991 in Russian translation in the USSR and will soon be published in China. His edited book *Science and the Soviet Social Order* was published in 1990.

Barbara Jancar-Webster, Professor of Political Science at the State University of New York in Brockport, works in the area of environmental policy in the countries of Eastern Europe and the USSR. Her recent publications in the field include several articles and the book *Environmental Management in the Soviet Union and Yugoslavia: Structure and Regulation in Federal Communist States* (1987).

Paul R. Josephson is a Professor at Sarah Lawrence College in the Science, Technology, and Society Department. His many publications center on the interrelations of science and politics in Russia and the

USSR. His book *Physics and Politics in Revolutionary Russia* will soon be published by the University of California Press.

John M. Kramer is Distinguished Professor of Political Science at Mary Washington College. He has also served as a Research Associate of the Russian Research Center at Harvard University and a Senior Fellow at the National Defense University. His numerous published works have focused on energy policy, the environment, and other social and political issues relating to technology in the USSR and Eastern Europe. His book *The Energy Gap in Eastern Europe* was published in 1990.

Susan J. Linz, Associate Professor of Economics at Michigan State University and editor of the journal *Comparative Economic Studies*, is a specialist on the Soviet economy who is currently working on innovation decision making, technological advance, and economic reform in the USSR. Editor of and contributor to *The Impact of World War II on the Soviet Union* (1985), *Reorganization and Reform in the Soviet Economy* (1988), and *The Soviet Economic Experiment* (1990), she has also published articles on Soviet management decision making.

Steven W. Popper, Full Economist at Rand, has done research on the technological effect of economic reforms in Eastern Europe. He is currently leading a multi-year project on U.S. policy toward Eastern Europe. Among his recent publications are Rand reports on *East European Reliance on Technology Imports from the West* (1988) and *Prospects for Modernizing Soviet Industry* (1990).

James P. Scanlan is Professor of Philosophy and Director of the Center for Slavic and East European Studies at the Ohio State University. A specialist in the history of Russian thought and contemporary Soviet ideology, he is the author of many works on those subjects, including the book *Marxism in the USSR* (Cornell, 1985). He also edits the quarterly journal *Soviet Studies in Philosophy*.

technology, culture, and development

The Experience of the Soviet Model

The Fits and Starts of Russian and Soviet Technology

Loren R. Graham

Many contemporary Western observers of the Soviet Union know that technological backwardness has been a major problem for that nation. These same observers often assume that the history of the development of technology in that country is a fairly recent one, at least compared to the leading Western nations. Some even believe that in the nineteenth century Russia was basically undeveloped and that only in the twentieth century did the technological gap between the Soviet Union and the Western nations begin to close.

A closer look at the history of technology in Russia and the Soviet Union reveals a much more complicated story. Instead of the picture of a steady bridging of a gap that begins in the twentieth century, we see a jagged curve stretching back several centuries, with individual high points of excellence achieved long before 1917, only to be lost again in succeeding years. And this repetitive story of momentary achievement followed by obsolescence continues into the twentieth century. The problem of technology in the history of Russia and the Soviet Union does not seem to be primarily one of the transfer of technology from other nations—something that has been going on there for centuries—but of sustaining the application of technology in the economy and constantly improving that technology. Technology is not something that, once obtained, develops on its own. It requires a culture and an economy that are receptive and stimulating. Creating such a culture and economy has been Russia's real problem in the fostering of technology.

Evidence of isolated early achievements in technology is abundant in Russian history. As early as the sixteenth century the casting technology of the Moscow Cannon Yard astonished Western visitors, for here were cast the largest church bell ever made and hundreds of heavy cannon for the Russian armies.[1] Among the bells and cannon produced some were highly decorated. Originally tutored by Western foundry-men, the Muscovites developed their own procedures, which they kept secret from foreign visitors. In 1632, the Dutchman Andrei Vinius established near Tula, south of Moscow, a forge for the manufacture of armaments that has had a continuous history to the present day. Peter the Great gave Russian technology, especially that which could be used for military or naval purposes, a great boost in the early eigh-teenth century. Not only did he import foreign technicians but he also sent Russian mechanics abroad for education. One of them, Andrei Nartov, became a master machinist who developed lathes, mint presses, guns, and locks.[2]

Many people are surprised to hear that at the end of the eighteenth century Russia was the largest exporter of iron in the world; its best customer in the iron trade was England.[3] In 1766 a Russian inventor, I. I. Polzunov, developed a 32-horsepower steam engine to pump water out of mines. Soviet historians have made much of Polzunov's engine, which preceded that of James Watt by several years and was evidently an improvement over the earlier Newcomen engine.[4] However, Polzunov's engine frequently broke down and was soon forgotten; its importance was not as a practical achievement, but as an instance of early Russian interest in steam technology.

An illustrative example of how Russian technology often advanced in spurts, aided by foreign assistance, only to fall behind later can be found in the Tula arms factories. As already mentioned, they were established by a Dutchman in the seventeenth century; at first they employed the most modern methods. By the time of Peter the Great in the early eighteenth century, however, they were lagging behind West European technology. Peter ordered their modernization and especially the greater utilization of water power. He brought in Swedish, Danish, and Prussian gunsmiths to teach Russian apprentices. After Peter's death, the policy of obtaining foreign assistance continued. Catherine the Great took an interest in the Tula factories in the last third of the eighteenth century,[5] actually helping to forge weapons during one visit. She also ordered that Russian gunsmiths be sent to England to improve their skills. During the Napoleonic Wars, the Tula factories were major suppliers of guns of various calibers to the Russian armies.

In order to illustrate more fully the cyclical nature of the history of Russian technology—the pattern of momentary excellence followed by obsolescence—I would like to delve more deeply into the history of the Tula armory in the first half of the nineteenth century. I will draw on unusually explicit primary sources that we possess on the conditions of that factory, and I will also compare the history of this factory with that of U.S. armories in the same period.

The tsarist government in the early nineteenth century was very proud of its army and the weapons that it disposed. That army was the largest in Europe, one million men strong, and the Russian Empire had proved, after the defeat of Napoleon, that it was the dominating military power on the Continent. In 1814 the Russians occupied Paris. Only British naval power counterbalanced the military forces of the tsar.

Realizing that this army must be equipped with weapons equal to those of its potential adversaries, the tsarist government made a major effort to modernize its armory in Tula in the immediate post–Napoleonic period. In 1817, a master English gunsmith, John Jones, was brought with his family to Tula to manufacture gun locks by means of dies instead of the previous method of manual forging. He also introduced the use of drop hammers and announced a program of producing interchangeable parts for his guns. By 1826 Jones had carried out such an impressive program of modernization that a tsarist inspector declared that the Tula factory "had been improved to such an extent that not a single weapon factory in the world can be compared with it."[6]

Wishing to observe personally the wonderful progress being reported in his Tula armory, Emperor Nicholas I decided to go to Tula on an official visit. He was told that at that time the armory had in its inventory 52,125 small arms produced by the new methods—a remarkable quantity. He was also told that no other country in the world was capable of producing such a large number of weapons with interchangeable parts.

Here are excerpts from the official report of Tsar Nicholas's visit of 1826:

> His Majesty the Emperor, accompanied by His Royal Highness Prince Carl of Prussia, . . . departed from Moscow for Tula on the morning of the 20th of September [1826]. . . . His Majesty traveled to the factory and remained there until six o'clock in the evening. . . . In the [gun]

lock assembly room, His Majesty the Emperor, wishing to make sure by experiment whether the interchangeability of the lock parts was now perfectly achieved in Tula or not [the realization of such an interchangeability was always considered to be impossible], took a few from a large number of finished locks and ordered them to be dismantled into parts. In the same way His Majesty ordered new locks to be assembled from them. These newly assembled locks had as free movement as if the parts of each lock were fitted to each other. This decisive test proves more than anything else the superiority of methods introduced, at the present time, by Mr. Jones for the manufacture of locks in the Tulskii factory. . . . Soon after, His Majesty for the second time deigned to visit the Tulskii factory. . . . In order to view in detail the different sections of the factory, the Great Prince stayed in the factory for two whole days. . . . In order to be certain about the perfect interchangeability of not only the lock parts, but also the barrels and even all the other parts of the gun without discrimination, His Majesty deigned to choose by himself thirteen finished guns in the armory and ordered them to be dismantled into parts. From these parts initially mixed he ordered assembly once more of complete new guns. All of the parts were fitted to each other with a perfect accuracy. This test proves beyond doubt that the gun industry in the Tulskii factory has been brought, under the High Management of His Imperial Majesty, to the highest degree of completion known at the present time.[7]

If this story is correct, it records a remarkable event. Historians of technology now agree that true interchangeability of parts for mass-produced small arms was not achieved elsewhere until the 1840s, when Americans in New England reached this goal.[8] Of course, many earlier claims had been made for interchangeability, dating back a half century. I will briefly mention some of those earlier claims a bit later, drawing on the work of my colleague Merritt Roe Smith and other historians of technology.

At this point the story becomes quite interesting and even paradoxical. We will probably never know exactly what happened during the tsar's visit to Tula in 1826, but evidence is mounting that he was deceived and that the guns manufactured in Tula at that time did not have truly interchangeable parts. The paradox, however, is that the evidence also indicates that the Tula factory in 1826 was, indeed, about as good as any large armory of the world. Some of its machines, such as the drop hammers and milling machines, were very impressive. The English master mechanic Jones, and perhaps some unknown Russian workmen, had made some improvements in Russia on the equipment that Jones had used in England.

What causes us to suspect that the tsar was deceived in 1826? And what evidence do we have that, despite this probable deceit, the Tula factories were approximately equal to the best armories in the world at that time, those in the United States and England? Historians of technology have recently examined surviving Russian small arms from the years 1812 to 1839 and have found evidence contradicting the Russian claims. The parts often bear the marks of hand filing, indicating that they were not interchangeable as originally produced, but instead had to be laboriously and expensively fitted by hand. Some of the parts are even numbered, a practice not necessary if true interchangeability has been achieved. The U.S. historian of technology Edwin Battison examined this evidence in 1981 and noted that the parts of the guns were no more interchangeable than most guns produced in the United States in the same years. He observed, "Similar parts on similar guns made for the United States service at the same times would show comparable marked parts."[9]

Battison then went on to ask:

> How is it possible that the Czar could have the incredible good fortune in his tests of locks and muskets? If such favorable results could be secured throughout the entire product of the factory by the workmen there would be no excuse for fitting the parts together before hardening them as was done. Some other explanation has to be considered. Out of thousands of muskets it would be possible to select a small number that might be useful in such a demonstration. To select and prepare such exceptional muskets would be expensive, to place them so the Czar could innocently choose them, apparently at random, would be highly deceitful but it may have been done.[10]

Before we conclude that this sort of deceit is essentially Russian, we should notice that similar deceits were evidently being practiced elsewhere at this time, most notably in the United States. In recent years U.S. historians of technology have destroyed the myth that Eli Whitney was the first developer of interchangeable parts.[11] In 1801, before a distinguished audience that included John Adams and Thomas Jefferson, Eli Whitney disassembled, intermixed, and reassembled the parts of ten gunlock mechanisms with no tool other than a screwdriver. Jefferson was so impressed that he wrote James Monroe that "Mr.

Whitney has invented moulds and machines for making all the pieces of his locks so exactly equal, that take 100 locks to pieces and mingle their parts and the hundred locks may be put together as well by taking the first pieces which come to hand."[12] Jefferson's enthusiasm was understandable; such guns could be easily repaired in the field.

We now know that Eli Whitney's claims were false. His muskets were not made of interchangeable parts. A historian who many years later studied the physical evidence concluded that the parts were "in some respects . . . not even approximately interchangeable."[13] Furthermore, Whitney was not able to achieve true interchangeability of parts during his lifetime, even though he remained a great exponent of the idea.

In the year 1826, one year after Whitney's death and the same year as the Tula demonstration, by remarkable coincidence, three reports were published evaluating the state of small-arms manufacture in the United States and Russia. They allow us to compare the methods of manufacture in the two countries quite accurately. The three reports were the Carrington Report on Harpers Ferry armory, an evaluation of Whitney's arms factory in Connecticut, and a report on the Tula armory in Russia.[14] The report on the Tula factory was the most detailed. The three reports, combined with physical evidence, show that guns with truly interchangeable parts were not being manufactured in large numbers in either country, but that Russia was probably equal to the United States at that time in most operations and was actually superior in its ability to produce very large numbers of modern guns. Battison observed that "comparison of the scant few machines in Whitney's possession at his death with the number and variety in use at Tula . . . certainly punctures and deflates the overblown popular folklore enveloping Whitney."[15] John Hall at Harper's Ferry had produced 2,000 breech-loading flintlock rifles on an innovative interchangeable system that was promising for the future, but the Russians were already producing more than 20,000 guns a year that were comparable to the best U.S. muskets of that time.

And yet, starting from a position of approximate equality in the manufacture of small arms in the 1820s, during the next thirty years Russia fell dramatically behind. From the 1830s through the 1850s, U.S. arms makers converted interchangeable-parts manufacture from an idea to a reality, and they expanded it into what became the "American System of Manufacture."[16] Russia missed out on this development.

The gradual slippage of the Russian Empire was concealed for a while by the fact that the wars that the Russians fought in the 1820s and the 1830s were against Turks and Caucasian mountaineers whose arms were inferior. The revelation of Russia's military obsolescence came at midcentury, when on its home soil of the Crimea the Russian army tried to resist vastly better-equipped British and French troops.

In the Crimean War the primary weapon of the Russian infantry was the smoothbore musket. Some of the muskets were even flintlocks, since the program of conversion to percussion muskets initiated in 1845 was still incomplete. Many of these weapons were in poor repair; their parts were not, by and large, replaceable. At the battle of Alma and Inkerman in the fall of 1854 the Russian troops faced French and British troops armed with rifles and minié balls, which had a lethal range approximately three times that of the Russian muskets. One of the Russian staff officers wrote that

> at the present time the greatest subject with us is stutsers [rifles]; we have powder, shot, troops, but they have not sent one stutser and evidently there is no hope of it. Seeing how in the Inkerman action, whole regiments melted from the stutsers, losing a fourth of their men, ... I am convinced that they will cut us down as soon as we fight in the open.[17]

One of the British officers said he shuddered "to see the devastation that the minié carbines produced in the ranks of the Russian columns," whose musket fire "did not carry half the distance to the enemy as they pressed forward."[18]

How do we account for this slippage in quality of small arms in only a few decades? Edwin Battison has posed two possible answers: (1) The deception of claiming interchangeable manufacture in 1826 may have "inhibited further modernization in a country so autocratic that once the monarch had certified success . . . there was no way to pursue further appropriations for continued progress."[19] (2) Russia's diplomatic representatives abroad failed to report on progress being made in armaments manufacture in other countries.

These answers may be partial explanations, but in my opinion they overlook the most important factor. In other countries, and particularly in the United States, the social and economic environment was one that nurtured and promoted technological development. It is the absence of such an environment in Russia that is most responsible for the difference in the rates of improvement of arms technology in Russia and Western nations. Without an environment that independently promoted innovation, modernization of Russian technology could come only at those moments when the tsarist government suddenly noticed slippage and ordered reforms on the basis of a new importation of Western experts and machines; such an abrupt rescue is what happened in 1817 when John Jones was brought to Tula, and is what happened after the Crimean War when the Russians again turned abroad for help with small arms. This time they paid special attention to American-made rifles produced by the methods of interchangeable-parts manufacture. Russian military technology, like all technology in Russia and later the Soviet Union, advanced in fits and starts.

It is very easy for me to observe that social and cultural factors played important roles in the failure of the Tula armories to keep up with Western developments, but can I make this hypothesis more persuasive? Do we have independent evidence of the influence of social factors on innovation in the field of weapons production in the first half of the nineteenth century?

A comparison of what was going on in the same years at two other national armories, those in the United States at Harpers Ferry, Virginia, and at Springfield, Massachusetts, may shed light on these questions. Merritt Roe Smith's 1977 book on the Harpers Ferry armory is particularly helpful here.[20] I will suggest that Tula fell behind the leading Western nations in the production of high-quality small arms for at least some of the same reasons that Harpers Ferry fell behind Springfield in the same years.

In his careful comparison of the Harpers Ferry and Springfield armories, Smith concluded, "Harpers Ferry remained a chronic trouble spot in the government's arsenal program. Tradition-bound and often recalcitrant, the civilian managers and labor force accommodated themselves to industrial civilization most uneasily. The situation at the Springfield armory presented a marked contrast. There factory masters as well as mechanics seemed to embrace the new technology without any of the hesitancy or trepidation of their contemporaries in Virginia."[21] Springfield, Smith continued, possessed an "expansive go-ahead vision," while attitudes at Harpers Ferry were "myopic in quality—at once restricted in scope, static in attitude, and provincial in its processes."[22] As a result, Harpers Ferry fell behind in innovation, while Springfield and the Connecticut River Valley were the birthplace of the new system of production, an approach that spread from arms manufacture to the rest of U.S. industry.

Smith found the roots of the difference in attitudes toward modernization in the contrasting social environments of Harpers Ferry and Springfield. Harpers Ferry was a small town in a predominantly rural area where a preindustrial ideology, craft traditions, and a social hierarchy influenced by slavery resisted the development of modern manufacturing methods. The town of Harpers Ferry was controlled by a few favored families who were suspicious of social change. These privileged families could provide education for their children, but for other families, educational opportunity was extremely limited. A Lancastrian school existed in Harpers Ferry from 1822 to 1837, but then failed for lack of public support.

The institution of slavery influenced social attitudes at the Harpers Ferry armory, even though few slaves were actually employed there. As Smith observed, "Because slaveholding heightened one's social standing in the community, the very possession of slaves by a few armorers made them extremely jealous of their rights and sensitive about the honor and dignity accorded their jobs. Any organizational or technical change that even slightly threatened to undermine their status was therefore firmly resisted. For instance, the most frequently repeated protests during the 'clock strike' of 1842 accused the Ordnance Department of subverting the worker's freedoms and making them 'slaves of machines.' In a patriarchal society both terms conveyed deeply felt meaning."[23]

Intensely proud of their special status as "master armorers," the veteran workers at Harpers Ferry resisted regimentation, machine work, time controls, and uniformity in production. In fact, when one superintendent named Thomas Dunn attempted in 1829 to enforce strict controls over work performance and quality of production, he was assassinated by one of the armorers. On other occasions the workers went on strike when they believed their status as craftsmen was being reduced to that of mere day laborers.

The Springfield armory presented a sharply contrasting picture in the same years. There the majority of the workers had come from farms and villages in western Massachusetts, where they benefitted from the public-education system. As former free farmers who had never known slavery, they did not fear loss of their social status.

Rather than fear the adoption of machinery, they greeted its advent. Reared in a controlled environment in which the Puritan ethic still had force, they readily accepted the regulated environment of factory life. Springfield's workers were, compared to those of Harpers Ferry, remarkably disciplined and industrious. The managers of the Springfield factories were aggressive and devoted to technological progress.

Against the background of this contemporaneous example of the way in which the social environment influenced receptivity to technological innovation in the United States, let us examine the social context of the Tula armory.[24]

By the first decades of the nineteenth century, the Tula arms factory was a complex of buildings and mills employing a great number of workers. It was, in fact, one of the largest armories in the world. In 1826, the Tula armory employed almost 14,000 workers, of whom more than 3,000 were members of a special estate (*soslovie*) recognized by the government and exempted from military service and from all taxes. Although technically serfs belonging to the state, these gunsmiths enjoyed remarkable privileges so long as they remained in their positions and obeyed the officers in charge of the armory. Some of them eventually came to own serfs of their own.

The Tula works also employed 3,000 or 4,000 serfs belonging to landowning nobility in the region. The gentry permitted these serfs to work off their estates as long as they paid a quitrent (*obrok*).

The history of the Tula ironworks is replete with conflicts between armorers and the rest of the peasantry in the Tula region. The armorers constantly struggled for special privileges that would distinguish them from the other peasants. Normally, all peasants were required to give their owners services in the form of labor or cash payments, and they were also required to pay taxes to the state and to serve in the army when called. The armorers of Tula gradually cast off most of these obligations. In their petitions to the tsar they reminded their ruler that they possessed special skills necessary for conducting wars, and they asked for freedom from obligations. These requests caused envy among the other peasants and sometimes open conflict. In order to avoid such confrontations, the tsarist government in the eighteenth century ordered the town of Tula to be split into different sections, with the armorers all living together in one settlement in which other citizens were forbidden to reside. The armorers did much of their work in their homes, and in peacetime they were allowed to produce tools, samovars, locks, and fittings, which they sold privately.

Despite the privileges that they enjoyed, the Tula armorers were also tightly regulated. Without the permission of the government they could not leave Tula or abandon their profession. If they fled, they could be forcibly returned. In 1824 when an armorer named Silin fled Tula, he was captured, brought back to the armory, and given two thousand strokes with a birch rod—surely an unsurvivable sentence. We are told by Soviet historians that in the eighteenth and nineteenth centuries there were more than two thousand attempted flights from the Tula armory, averaging one a month.[25]

Soviet historians have also emphasized that the Tula workers were economically exploited and suffered from poor living standards. Closer examination, however, seems to indicate that peasants and workers in the Tula region lived rather well compared with other peasants in the empire. One Western historian who studied serfs in the neighboring province of Tambov and made some comparisons with the Tula region maintained that "Russian serfs simply were not poorer than peasants elsewhere."[26] But this same historian also observed that "Serfdom was, of course, economically exploitative, but it was far more socially oppressive than economically onrous." Life among the peasants in the region, he continued, was "hostile, violent, vengeful, quarrelsome, fearful, and vituperative."[27] The control of the owners over the serfs was bad enough, but even worse was the "social oppression of serf over serf," which secured the system of authority. In Tula when armorers became serf owners or serf supervisors, which many of the master armorers did, they soon gained reputations as extremely severe taskmasters.

The most prestigious workers in the Tula armory were the master armorers who produced richly ornamented guns. The master armorers specialized in such techniques as sunken relief, damascening, inlaying, chasing, blueing, and the carving of hunting scenes on the stocks of the guns. These armorers considered themselves artists, not workers, and some of them, such as Petr Goltiakov, became famous. Goltiakov started out as an ordinary gunsmith, but as a result of his outstanding talents, he became lock overseer of the entire armory; in 1852, just before the outbreak of the Crimean War, he was appointed purveyor of guns to the Grand Princes Nikolai and Mikhail, a position of great honor.[28] Goltiakov's guns were "presentation arms" of the type made for the ruling family and the highest officers, not the ones made for the infantry. The system of rewards for armorers at Tula was heavily skewed in favor of craftsmen like Goltiakov producing presentation weapons, not machinists making ordinary arms. As a result, a few of the best Russian guns were very good, but the average were of poor quality.

The master craftsmen of Tula relied on their personal skills, and they resisted any innovation that would reduce their status to the ranks of state peasants, where most of their families started. Tula had a long history of resistance to machines.[29] During the time of Peter the Great, the master armorers submitted a protest to the Senate against the introduction of water power. For decades the armorers resisted the transfer of work from their homes, where many of them had small forges, to centralized plants. In 1815 the tsarist government installed a steam engine in one of the buildings in Tula, but the official report of 1826 to which I have referred tells us that the engine was still not being used. At that time most of the armorers worked at home at hand labor rather than in the main shops. Soviet historians tell us that even as late as 1860 only 35 percent of the Tula armorers worked "within the walls of the factory" rather than at home.[30]

In 1851 the Crystal Palace Exhibition was held in London, a grand affair where many nations displayed products of art and industry. The Russians, along with Austrians and the French, presented artistic displays in which the artifacts were heavily ornamented. The U.S. display was strikingly different. As the historian Nathan Rosenberg observed, "A visitor who came to the American exhibit to gratify his aesthetic sensibilities was wasting his time. It was severe and utilitarian in nature."[31] Critics soon called the U.S. exhibition area the "prairie ground." Among the items on display were ice-making machines, corn-husk mattresses, railroad switches, and telegraph instruments. More to the point, the display included small arms produced by the firms of Samuel Colt of Connecticut and Robbins and Lawrence of Vermont. The R. & L. weapons were made of truly interchangeable parts. Colt announced that he relied on machine production because "with hand labour it was not possible to obtain that amount of uniformity, or accuracy in the several parts, which is so desirable."[32] This 1851 display was a warning for the Tula armorers. The future was one in which art was to be lost, but lethal uniformity was to be gained.

The story of early achievement and subsequent decline that we saw at the Tula arms works is a common cyclical pattern in the history of Russian and Soviet technology. I will mention a few other similar cycles. Although in 1700 Russia was the largest producer of iron in the world, by the first decades of the nineteenth century the English iron industry was growing at twelve times the rate of the Russian industry, and by 1850 Russia ranked eighth among the nations of the world in iron production. By 1860 England was producing 4,000,000 tons of pig iron a year, while Russia produced less than 350,000 tons.[33]

In railroads Russia also showed early prominence, only to drop behind later. A Russian father-and-son team, the Cherepanovs, produced in 1835 a steam locomotive that could pull a sixty-ton load. It is possible that Cherepanov senior got his original idea about locomotives from the Englishman George Stephenson, since Cherepanov visited England at the time Stephenson was making locomotives. The Cherepanov locomotive was not developed further, however, and Russia soon had to turn to foreigners, and particularly Americans, for the construction of railways. Nonetheless, Russia was one of the handful of pioneering nations in the construction of railways. The first steam railway in Russia was opened to the public in 1837, the same year as the first one in Austria and only five years after the first steam railway in France.[34] St. Petersburg and Moscow were connected by rail before New York and Chicago.

In the 1840s the U.S. engineer Joseph Harrison, Jr., supervised the building in St. Petersburg of what was at that time one of the world's most impressive railroad factories. In 1846 the Aleksandrovskii railway works in St. Petersburg employed 1,920 workers, 85 percent of whom were Russian peasants. The factory produced ten freight cars a day, four passenger cars a month, and six locomotives and tenders a month. In 1847 Harrison exulted in a letter to his father-in-law, "We shall have all the Locomotives done in three Months from the present time at the rate we are now making them—eight a month or two in six working days.... This beats our old operations at home," he continued, "and it is about two engines a month ahead of any establishment in any country that we have any knowledge of."[35] For a few years foreign observers agreed that the tsarist empire possessed the largest and best-equipped railways in the world.

Russia's prominence in railroads did not last long, however. The railway network in Russia expanded much more slowly than in Britain, France, Germany, or the United States. By 1855 Russia possessed only 653 miles of railway, compared to 17,398 in the United States and 8,054 in England.[36]

In the development of electricity a similar pattern emerged of an early promising start followed by a slump. When the streets and public gardens of Paris and London were first electrified in the late 1870s and 1880s, the method of illumination was the arc light, patented in Paris in 1876 by the Russian inventor Pavel Iablochkov. At the 1878 World Exposition in Paris, Iablochkov's lights were exhibited to great acclaim. The new street lamps were popularly referred to as "Russian lights."[37] Impressed by his success in Western Europe, Iablochkov returned to Russia and attempted to manufacture and sell his lights there. Without an adequate market at home and unable to innovate rapidly enough to keep up with foreign firms, Iablochkov's company failed miserably.[38] The major cities of the Russian Empire were eventually electrified by foreigners. Electrification of the countryside proceeded very slowly and was a major task of the Soviet government for decades after the 1917 Revolution.

In the last decades of the tsarist empire further examples of the alternation of periods of technological achievement followed by retardation continued to appear. For example, in 1900 Russia was the world's largest producer of crude oil. In 1912 a plant in Moscow began producing a domestic automobile in small numbers. Igor Sikorsky even designed and built a four-engine airplane in 1913; more than fifty of these large planes were used as bombers in World War I. But in backward tsarist Russia there was no possibility of a mass market for automobiles and little chance for development of an aeronautical industry, outside of military needs. Petroleum was used primarily to produce kerosene for illumination and heat, not for gasoline. When popular demand for automobiles in other countries, particularly the United States, caused a dramatic jump in the need for gasoline, the center of production of petroleum products moved elsewhere.

The pattern of cycles of innovation and stagnation in technology continued into the Soviet period. When the Soviet Union rapidly industrialized in the late twenties and thirties, it adopted the latest technology from abroad. At various moments the Soviet government granted concessions to foreign concerns, formed joint-stock companies, hired foreign consultants, or simply illegally copied Western technology. In the period from 1929 to 1945 about 175 technical assistance agreements were arranged between the Soviet Union and Western companies; the latter included the best and largest firms in the world: Ford, International Harvester, Krupp, Pennsylvania Railroad, Pratt and Whitney, Siemens, Standard Oil, Union Oil Products, Babcock and Wilcox, Bucyrus Erie, Caterpillar Tractor, Dupont, Metropolitan-Vickers, and many others.[39]

Some Western observers believed that by modernizing after other industrial nations the Soviet Union would benefit from its ability to pick and choose the latest technology. Professor Alexander Gerschenkron wrote a series of famous articles in which he spoke of "the advantages of backwardness," noticing that the Soviet steel and automobile industries were being constructed on the very latest Western models.[40] The great steel plant of Magnitogorsk was consciously modeled on the United States Steel plant in Gary, Indiana, where the leading Soviet metallurgical engineer, Ivan Bardin, had once worked. Gary at that time represented the last word in steel making. The large auto plant in the city of Gor'kii was built by engineers and workers from the Ford Motor Company and was modeled on the River Rouge plant in Detroit, considered to be the most modern, complete auto plant in the world.

Once Soviet factory directors achieved independence in production, they tried to shed the foreign tutelage. Soviet leaders believed that, after industrial parity had been achieved, what they saw as the inherent superiority of a socialist economy would then lead to further qualitative improvements and technological originality. Time demonstrated, however, that they overestimated the economic efficiency of socialism and they underestimated the deadening effects on technology of their social and political backwardness.

Once basic industrialization had been attained by the early sixties, the debilitating effects of a state-owned and centralized economy and the absence of a sustaining environment showed up ever more clearly. Although the Soviet Union became the world's largest producer of many basic industrial products, including oil (once again), steel, cement, and machine tools, it offered the world market very little in terms of original or qualitatively superior products. And even quantitatively it began in the seventies to slip in relative terms. After becoming the second-largest economy in the world, surpassed only by the United States, the rate of industrial growth in the Soviet Union sagged dramatically. This failure of the Soviet economy to continue to grow and to innovate was a major cause of the economic and political reforms initiated by Mikhail Gorbachev in the last half of the eighties. Those reforms included a turn once again to Western companies for technological assistance by means of joint ventures and other agreements.

After Gorbachev's rise to power in the mid-eighties, engineers from USX, the descendant of U.S. Steel, and the Ford Motor Company were asked to come to the USSR and to examine the possibility of modernizing the Magnitogorsk and Gor'kii plants that the U.S. companies had originally helped build. In both instances, the U.S. engineers found that surprisingly little had changed since the original construction of the Soviet plants. Magnitogorsk had become the rust belt of the Soviet Union, a monument to the inefficient production of steel. By the eighties, U.S. steel plants were not doing so well either, but it just so happened that Gary, the old model for Magnitogorsk, was one of the few U.S. steel plants to be thoroughly modernized in the intervening years. At Gor'kii, the Ford engineers found that some of the original equipment of the thirties had been used as late as the seventies, and that the basic layout of the plant and its management principles were those of the River Rouge plant of the thirties. Several Ford engineers referred to the Gor'kii plant as a museum to the art of automobile manufacturing. The Ford Motor Company eventually declined the invitation to form a joint venture for the modernization of Gor'kii, partly because of the formidable nature of the task, but even more, it seems, because of their worries about being able to extract their profits.

As another example of the fits and starts of Soviet technology, one might notice that in the sixties the Soviet Union produced a fairly successful second generation computer, the BESM–6.[41] Designed by S. A. Lebedev, the BESM–6 operated at 10 megahertz and was rated at 1 MIPS (millions of instructions per second). It was a purely indigenous development and possessed a number of impressive features. While it was not the best computer in the world at that time, it was much closer to the technological level of the leading models in the West than current Soviet computers are. In other words, after a promising start, the Soviet computer industry has fallen further and further behind. Failing to keep up with the West on the basis of native Soviet designs, Soviet computer scientists canceled the BESM series and shifted to the RIAD series, which was a belated copy of the IBM 360 series.

A similar pattern can be seen in the Soviet nuclear-power program. The USSR was the first country in the world to produce a usable nuclear power plant, and it rapidly expanded the use of nuclear energy. Soviet engineers failed, however, to keep up with the advances in instrumentation, reliability, and safety procedures developed in other countries. The disaster at Chernobyl in 1986 was a shock to both the Soviet public and the Soviet nuclear-power industry. In the late eighties many planned Soviet nuclear power plants were canceled. Western experts who visited Soviet nuclear power plants were alarmed by the potential for accidents.

An area of technology where the Soviet Union obviously made spectacular gains in the last half of the twentieth century was space exploration. The USSR was the first country in the world to launch an artificial satellite, first to launch a person into space, and first to perform a "space walk" in which a cosmonaut left his space vehicle. And although the United States was the first to send astronauts to the moon, in the seventies and eighties, the Soviet Union continued a very active and impressive space program.

The Soviet space program illustrates both the strengths and the weaknesses of Soviet technology. It was a centralized effort to which the government assigned first priority in obtaining talent and necessary materials. In its early phases it was heavily dependent on military technology, using knowledge gained in rocket development for intercontinental ballistic missiles. These are the sorts of activities in which the Soviet system has traditionally done well, in contrast to development of technology dependent on the decentralized civilian and consumer economies.

Russia possessed one of the great pioneers in the conceptualization of space travel in the figure of Konstantin Tsiolkovskii (1857–1935), an autodidact who gained recognition for his work only late in life. Tsiolkovskii elaborated a theory of multistage rockets, and he also proposed using clusters of rockets to achieve great speeds. He explored the mathematical relationship between the velocity of a rocket at any one moment, the velocity of the gas particles expelled from the nozzle of the rocket, the mass of the rocket, and the mass of the expended fuel. In 1897 he constructed the first wind tunnel in Russia, and he conducted a number of experiments with models of airfoils. However, his main achievement was the advancement of the idea of space travel and the derivation of basic principles rather than the actual designing and building of working rockets.

Looking closely at the early history of the Soviet space program, one is filled with admiration for its early leader Sergei Korolev, the "Chief Designer," whose name was kept secret until his death in 1966.[42] Korolev performed miracles in fulfilling the demands of the government under incredibly difficult conditions. Arrested in 1937 and thrown into one of Stalin's labor camps, Korolev worked on rocket technology for many years in a special prison laboratory, or *sharashka,* of the type described so vividly in Aleksandr Solzhenitsyn's novel *The First Circle.* After Stalin's death in 1953, Korolev was rehabilitated and drawn into work on military missiles. Ordered to develop a rocket of sufficient power and range to reach the United States, Korolev was confronted with the problem that the large rocket engines necessary for this task produced temperatures in the walls of the nozzles greater than any Soviet alloys could withstand. Special heat-resistant alloys were used in large U.S. rocket engines such as the "Atlas" and the "Saturn" that were unavailable to Korolev. Consequently, he adopted a dramatically different approach. He clustered smaller rocket engines in pods of four or five. The rocket that sent up the world's first satellite in 1957 had a four-chamber cluster. The rocket that sent into orbit the first man, Iurii Gagarin, was a giant "cluster of clusters" rocket with a total of twenty engines. Making all these engines and their associated fuel pumps and systems work simultaneously was an engineering accomplishment of the first magnitude. It was not the most direct and efficient solution of the problem, but it worked.

No sooner would Korolev fulfill one of Khrushchev's demands for a space spectacular than the Soviet leader would present him with another. The most extreme of these requests for presentation technology was undoubtedly one in 1963 that the Soviet Union launch three men in one space vehicle before the United States succeeded in launching two in a single capsule. To carry out that order Korolev had to choose cosmonauts of small stature, to ask them to drop the precaution of wearing bulky space suits, and to pack them so tightly in a small sphere that they were arranged around each other like pretzels. But the effort succeeded, and on time.[43]

After Khrushchev's overthrow in 1964 and after the United States surpassed the Soviet Union with its successful landing of astronauts on the moon in 1969, the Soviet space program became less hectic. Soviet authorities maintained that they had never been engaged in a race with the United States to put a man on the moon, but in 1990 a group of U.S. aeronautical engineers from the Massachusetts Institute of Technology was shown an old Soviet lunar lander intended for that purpose and were told that the Soviet Union only abandoned the race to the moon when it was clear the Americans would be first. In the seventies and eighties the Soviet space program made steady progress at a time when the U.S. program was uneven. By the late eighties and early nineties, however, the Soviet space program encountered increasing criticism from a now vocal public, which saw it as a drain on resources needed in the domestic economy.

What we see in the history of Russian and Soviet technology, then, is a pattern of fits and starts, repetitive cycles characterized by early achievement and subsequent failure. These cycles have occurred so often that we must look for a set of underlying causes. While the explanations for the cycles will be somewhat different in each case, I strongly suspect that social and economic barriers, rather than technical ineptitude, will be the common explanation. It is even worth noticing that a number of the technology-retarding characteristics of the society of the Tula armory in the nineteenth century continued to exist in the Soviet Union until quite recently, and in fact some still do. Workers in Tula in the nineteenth century were tied to their place of residence and were subject to strict regulation, just as workers were in Stalin's Soviet Union. (In fact, the system of *propiska,* or enforced residence permits, still exists in the Soviet Union.) Both in Tula and in the Soviet Union there was no free market and no system of competitive bidding by private contractors. In tsarist Russia and the Soviet Union, social hierarchies reigned in the towns and the workplaces. Under the tsars and under Communist rule, it was more important to please the political authorities by announcing spectacular and showy achievements (what I have called "presentation technology") than it was to be efficient or cost-effective. And living standards in both societies have depended more on rank and access to influential people and institutional services than on achievements or salaries. These conditions created, in both tsarist and Soviet Russia, a society where the inertial forces were enormous. A graphic description of these inertial forces was given in fictional form many years ago by Vladimir Dudintsev in his novel *Not by Bread Alone,* in which a lone inventor of a new means of producing steel pipes vainly fights the Soviet bureaucracy. Dudintsev was silenced by the censors of Brezhnev's Soviet Union, but under Gorbachev he has emerged with a new novel describing resistance to scientific and technological innovation.

I will conclude by observing that Gorbachev realized that the technical modernization of the Soviet Union cannot be achieved by transferring technology from the West while attempting to hold the social, political, and economic system constant. That was the approach of the tsars, of Stalin (after the first five-year plan), and of Brezhnev; it failed. The cycle that Gorbachev wanted to break out of was the old repetitive one of early achievement, based on native talent and government priorities, followed by retardation because of social resistance and the lack of economic stimulation. Gorbachev understood that this cycle could only be broken by the transformation of the society itself.

Notes

1. Edward V. Williams, *The Bells of Russia: History and Technology* (Princ.. 'on: Princeton University Press, 1985).
2. V. S. Virginskii, *Tvortsy novoi tekhniki v krepostnoi Rossii* (Moscow: Gosudarstvennoe uchebno-pedagogicheskoe izd., 1962); F. N. Zagorskii, *Andrei Konstantinovich Nartov, 1694–1756* (Leningrad: Nauka, 1969); and M. E. Gize, *Nartov v Peterburge* (Leningrad: Lenizdat, 1988).
3. Arcadius Kahan, *The Plow, the Hammer, and the Knout: An Economic History of Eighteenth Century Russia* (Chicago: University of Chicago Press, 1985).
4. V. V. Danilevskii, *Russkaia tekhnika,* 2d ed. (Leningrad: Gazetno-zhurnal'noe i knizhnoe izd., 1948).
5. *Istoriia Tul'skogo oruzheinogo zavoda, 1712–1972* (Moscow: Mysl', 1973).
6. Iosif Gamel, *Description of the Tula Weapon Factory in Regard to Historical and Technical Aspects,* ed. E. A. Battison (New Delhi: Amerind Publishing, 1988), p. 1.
7. Ibid., pp. 6–8.
8. Merritt Roe Smith, *Harpers Ferry Armory and the New Technology: The Challenge of Change* (Ithaca: Cornell University Press, 1977), p. 325 and *passim.*
9. Edwin A. Battison, "Introduction to the English Edition," in Gamel, *Description of the Tula Weapon Factory,* p. xxiv.
10. Ibid., p. xxii.
11. Merritt Roe Smith, "Eli Whitney and the American System of Manufacturing," in Carroll Pursell, Jr., *Technology in America: A History of Individuals and Ideas* (Cambridge: MIT Press, 1981), pp. 45–61.
12. Ibid., p. 47.
13. Ibid., p. 48.
14. See Battison, "Introduction," p. xii, in Gamel, *Description of the Tula Weapon Factory*; and James Carrington et al., "Examination of Hall's Machinery," manuscript of January 6, 1827, in *A Collection of Annual Reports, Chief of Ordnance,* vol. 1 (1812–44) (Washington, DC, 1878).
15. Battison, "Introduction," p. xii.
16. Nathan Rosenberg, ed., *The American System of Manufactures* (Edinburgh: Edinburgh University Press, 1969); Charles H. Fitch, "Report on the Manufactures of Interchangeable Mechanism," *Tenth Census of the U.S.: Manufactures* II

(Washington: 1883), pp. 611–45; and Joseph Wickham Roe, *English and American Tool Builders* (New Haven: Yale University Press, 1916).

17. John Sheldon Curtiss, *The Russian Army under Nicholas I, 1825–1855* (Durham: Duke University Press, 1965), p. 127.

18. Ibid.

19. Battison, "Introduction," p. xxv.

20. Smith, *Harpers Ferry Armory and the New Technology: The Challenge of Change* (Ithaca: Cornell University Press, 1977).

21. Ibid., p. 323.

22. Ibid., p. 330.

23. Ibid., pp. 333–34.

24. Gamel, *Description of the Tula Weapon Factory*; *Istoriia Tul'skogo oruzheinogo zavoda, 1712–1972*; V. N. Ashurkov, *Gorod masterov* (Tula: Tul'skoe knizhnoe izd., 1958); M. I. Rostovtsev, *Tula* (Tula: Tul'skoe knizhnoe izd., 1958); V. Mel'shiian, *Tula: ekonomiko-geograficheskii ocherk* (Tula: Priok. knizhnoe izd., 1968); and V. Berman, ed., *Masterpieces of Tula Gun-Makers* (Moscow: Planeta, 1981).

25. *Istoriia Tul'skogo oruzheinogo zavoda, 1712–1972*, p. 57.

26. Steven L. Hoch, *Serfdom and Social Control in Russia* (Chicago: University of Chicago Press, 1986), p. 187.

27. Ibid., p. 189.

28. Berman, *Masterpieces of Tula Gun-Makers*, p. 11.

29. *Istoriia Tul'skogo oruzheinogo zavoda, 1712–1972*, pp. 32–35.

30. Ibid., p. 52.

31. Rosenberg, *American System of Manufactures*, p. 7.

32. Ibid., p. 16.

33. Jerome Blum, *Lord and Peasant in Russia from the Ninth to the Nineteenth Century* (Princeton: Princeton University Press, 1961), p. 295.

34. Richard M. Haywood, *The Beginning of Railway Development in Russia and the Reign of Nicholas I, 1835–1842* (Durham: Duke University Press, 1969), p. xvii.

35. Merritt Roe Smith, "Becoming Engineers" (unpublished ms. of 31 August 1987), p. 32.

36. Haywood, *Beginning of Railway Development*, p. 242.

37. L. D. Bel'kind, *Pavel Nikolaevich Iablochkov, 1847–1894* (Moscow: Izd. Akademii nauk SSSR, 1962); M. A. Shatelen, *Russkie elektrotekhniki XIX veka* (Moscow-Leningrad: 1955).

38. Jonathan Coopersmith, "The Role of the Military in the Electrification of Russia, 1870–1890," in Everett Mendelsohn, Merritt Roe Smith, and Peter Weingart, eds., *Science, Technology and the Military*, vol. 2 (Dordrecht: Kluewer Academic Publishers, 1988).

39. Antony Sutton, *Western Technology and Soviet Economic Development, 1917–1965*, vols. 1–3, (Stanford: Hoover Institution Press, 1968, 1971, 1973); and especially vol. 2 (1971), pp. 363–72.

40. Alexander Gerschenkron, *Economic Backwardness in Historical Perspective* (Cambridge: Harvard University Press, 1962).

41. Gregory Crowe, Department of the History of Science, Harvard University, is writing a history of Soviet computing.

42. On the Soviet space program, see James E. Oberg, *Red Star in Orbit* (New York: Random House, 1981); Nicholas Daniloff, *The Kremlin and the Cosmos* (New York: Alfred A. Knopf, 1972); Walter A. McDougall, *The Heavens and the Earth: A Political History of the Space Age* (New York: Basic Books, 1985); Leonid Vladimirov, *Russian Space Bluff* (London: Tom Stacey, 1971); and *Uspekhi SSSR v issledovanii kosmicheskogo prostranstva* (Moscow: Nauka, 1968).

43. Oberg, *Red Star,* pp. 74–77; Vladimirov, *Russian Space Bluff.*

Science and Technology as Panacea in Gorbachev's Russia

Paul R. Josephson

What is the role of modern technology in a society undergoing rapid social, economic, and political change? Many social commentators have embraced modern science and technology as a panacea for socio-economic problems. The Taylorists and their Soviet counterparts believed in the power of time-motion studies to mediate all disputes between workers and managers. Representatives of various technocratic movements insisted that scientists and engineers armed with all of the "facts" could provide policymakers value-free answers to use in choosing the "one best way" to move society along. The Stalinist ideologue claimed that "Technology will decide everything!" while his Brezhnevite counterpart had unending faith in the infallibility of scientists and the fruits of their labor to harness the advantages of the developed socialist system to the scientific-technological revolution.

This essay examines the impact of the ongoing political and economic reforms in the USSR on Soviet attitudes toward technology. It considers the interaction of expert advice, politics, culture, and technology. In the first part, I look at the changing relationship between scientists, their professional organizations, the Academy of Sciences, and party and government institutions. I then turn to a discussion of several technologies. My conclusions are based on analysis of interviews; daily and weekly newspapers; literary, scientific-popular, and scientific weeklies and journals; and independent political sheets from Russia (Moscow, Leningrad, Siberia) and the Ukraine. I have tried to paint a picture of attitudes toward technology as perceived through scientists' and engineers' eyes.

Ongoing debates on the place of technology and big science—

nuclear and space research, computers, biomedicine, and ecology—in the USSR indicate that for many scientists and engineers technology is no longer the panacea it was during the Stalin, Khrushchev, and Brezhnev years. In some circles, big science continues to serve as a cure-all for Soviet socioeconomic problems, while in others, scholars have turned to the computer, market mechanisms, and the U.S. system as new saviors of the Soviet Union. At the center of these positions is the belief that technology is value neutral. Scholars hesitate to recognize science and technology as products of social, political, and economic forces. They reject the argument that technology is inherently political, requiring the creation of specific infrastructures and social relations for its introduction. Rather, they argue that technology can be applied for either constructive or destructive purposes; it can be used or abused in any social or political setting.

The physicists are the least inclined to indict science and technology for past excesses, a position they often share with conservative Party members. They are frustrated by the unwillingness of the public and government to give them the respect and support they once commanded. By all accounts, they wish to be freed from constraints imposed on their research by a naive and fickle public. They have maintained a kind of technological arrogance that has long characterized their approach to nuclear physics. "Technological arrogance" I define as support for technologies, often of limited social utility or questionable technical feasibility, whose introduction leads to an underestimation of environmental or social costs.

Most biologists and ecologists in the USSR, while more critical of the excessive support given to technological development in past periods of Soviet history, still adhere to a view of technology as neutral. They propose the creation of a technology-assessment process that would rely on institutions and analytical tools like those in the West, and would put a premium on the input of special expertise. The quantification of benefits and risks would, however, tend to exclude value systems and ethical and aesthetic considerations, while favoring extensive development projects over smaller ones by emphasizing aggregate data and economies of scale. Since life scientists desire a more circumspect approach to the evaluation of technology, not its rejection, they too seem to view technology as fitting the "use-abuse" categorization. They also fear social control of their disciplines, recalling the Lysenko years and the political interference in genetics.

On the other hand, Luddism of sorts also appears to be on the rise in the USSR. Among modern-day Luddites, technology is an evil. Some Soviet Luddites believe that it is possible to turn back the clock of modern technology and return to an agrarian-based economy. Other Luddites share the view of physicists and life scientists that modern societies are bound to build technology. They have a determinist view of technology as self-augmenting and autonomous. They wish not to destroy new technologies but to create a system in which it is easier to derail them. These attitudes are most prominent among such literary figures as Sergei Zalygin of *Novyi mir* and Valentin Rasputin. They condemn gigantomania. Their pastoral vision of unspoiled nature leads to rejection of all geological engineering projects—river diversion, dams, hydropower stations, nuclear power. The public seems to share this view of technology as inherently evil, but also as something beyond their control.

Soviet Marxism and Technology

The view of science and technology as a panacea is grounded in Soviet Marxism and reinforced by a fascination with U.S. technology that dates back to the first days of the Russian revolution. G. A. Cohen argues that historical materialism is a technologically determinist doctrine. This means that the development of the productive forces (machines, tools, instruments, science, technology, and people themselves) is the single most important factor in the course of human history, and further that science and technology provide the key to the socialist future.[1] Some have disputed Cohen's contention that for Marx "machines make history."[2] But whether Marx so argued does not obscure the fact that Soviet theorists and political leaders from Lenin, Bukharin, and Trotskii to Stalin, Khrushchev, and Brezhnev have embraced technological determinism. They have seen technology as "the highest form of culture," emphasized the development of the productive forces in the creation of communism, declared that "technology decides everything," and put their faith in the so-called scientific-technological revolution and the transformation of science into a direct productive force to achieve political, economic, and social goals.[3]

Tied to the Soviet version of technological determinism is a fascination with things American. Until Stalin's rise to power and the creation of autarkic economic and scientific relations with the West, Party figures, scientists, and engineers who traveled to the West reveled in the glow of U.S. technology—its skyscrapers, public transportation, automobiles and highways, and its industrial laboratories. They reported their findings in pamphlets, books, and such journals as *Priroda, Nauchnyi rabotnik, Bol' shevik,* and *Nauchnoe slovo.* The U.S. system of manufacture, its factories, *Fordizm,* and mass production in particular were seen as the key to the Soviet future. Even Il'f and Petrov, while critical of some aspects of American life, regaled the Soviet reader with fantastic tales of technology in the United States in *Odnoetazhnaia Amerika* (1936).[4]

Amerikanizm faded under Stalin during the construction of "socialism in one country." Bourgeois science and technology were seen as exploitative, serving the profit motive, not the masses. The Industrial party and Shakhty affairs signaled a class war against the engineer and what he stood for: "ivory tower reasoning," not practical applications. And yet, faith in technology did not abate. It would secure success in the construction of communism. Its display value—not merely its physical presence but also its ideological significance—grew at Magnitogorsk, Dneprostroi, and in the machine tractor station (MTS).[5] Faith in technology reached a high point in the Stalinist skyscraper, the construction of the metro and canals, and other projects infamous for their gigantomania, if not aesthetics.

After Stalin's death, the development of a "cult of science" put scientists and engineers at the forefront of the effort to use big science and technology to solve Soviet economic problems. The Soviet fascination with big science and technology grew unchallenged in the fifties on the foundation of successes in space, nuclear power, and high-energy physics, which rivaled those in the West. While some public displeasure with expenditures on technologies with great display value but limited significance for the consumer can be identified, by and large this opposition was muted until Gorbachev came to power, when glasnost and perestroika triggered a reevaluation of the place of technology in society.

Technology, Experts, and the
Public under Gorbachev

Glasnost created greater awareness of the potential social costs and environmental dangers of unregulated science and technology among scientists and lay people. Discussion of the Chernobyl disaster, failure of the Phobos 1 and 2 Mars probes, publicity over the recently scuttled plan to divert the flow of Siberian rivers from north to south, and previously suppressed reports of accidents with loss of life have all helped to call into question support for large-scale technology projects and big science.

Glasnost has also led to an examination of the excesses of Stalinism and the evils of the "command-administrative system," which was developed under Brezhnev. Nearly every month, such journals as *Priroda, Nauka i zhizn', Khimiia i zhizn'*, and *Vestnik Akademii nauk SSSR* run exposés on the purges, the death and suffering of innocents, and the poor performance of scientific R&D, which indirectly blame unquestioned faith in technology. Dailies and weeklies chronicle the environmental destruction wrought by profligate use of natural resources and intensive economic development—the ruin of rivers and lakes; the poisoning of land by pesticides; air pollution from outmoded factories and obsolete automobile engines; an impending disaster from AIDS because of the failure of the Ministry of Health to respond with resolve, a health-care system known more for the access it gives to the elite than its ability to reduce infant mortality and age-specific death rates; and astronomical expenditures on big science and technology (e.g., the Baikal-Amur Magistral [BAM], space, nuclear power) when environmental and medical issues are more pressing.

Finally, perestroika and glasnost have led to a reevaluation of the relationship between fundamental science, applied science, and technology and the roles of the Academy of Sciences, industrial ministries, and government. This is especially true in view of the "federalization" of the empire, the establishment of several alternative scholars' organizations, and the question of the proper role of the Supreme Soviet in light of the Communist party's declining role in forming science and technology policy.[6]

Soviet scientists and engineers do not question the power of their disciplines to contribute to the solution of the many problems facing the USSR today. Gone, perhaps, is some of the technological arrogance that characterized the nuclear-engineering profession before Chernobyl. But scientists on the whole continue to believe that scientific activity is value free. Hence, given the right political system and social foundations, and freedom to work without interference, their research naturally would produce great benefits for society. Some social scientists call for evaluation of the social and political context of modern technology to assess its impact at the earliest possible stage of development. Biologists criticize only the old institutions of science and technology, suggesting alternatives to the Academy of Sciences and various state committees, and rarely acknowledge the potential danger of unrestrained scientific research and technological development. Underlying all of these views is the belief that fundamental research must be given the freedom (and wherewithal) to develop without constraints; any regulation must come from within the scientific community since political and social interference smack of the "command-administrative system."

The first response of scientists and engineers to perestroika was democratization of management and decentralization of organization in existing scientific research institutes. More recently, there has been a resurgence of independent professional organizations of scientific experts. All independent professional organizations—architects, lawyers and engineers, biologists and physicists—were disbanded by the Party in the thirties. The Party feared independent loci of power, especially among groups with special expertise, and brought all of its weapons of control to bear, including coercion, purge, arrest, and execution. Now, however, professional organizations have been formed that represent all disciplines. The formation of these independent organizations indicates that scientists and engineers believe expertise free of social and political control will provide the best solutions to technological problems.

Such organizations as the Physics Society of the USSR, the Union of Scholars of the USSR, the Nuclear Society, and the Chernobyl Society were formed to help the government formulate science policy with three goals in mind: to defend specialists' professional interests; to provide the government with independent expertise to ensure well-informed policy decisions; and to combat incipient anti-science attitudes among the Soviet population. In other republics—Estonia, Latvia, and Belorussia—professional societies also were created.[7]

Another organization that promotes a major reevaluation of science and technology policy is the Union of Scholars of the USSR. Its members intend it to supplant the Academy of Sciences. The union lacks financial support and physical space, but attracts an increasing number of specialists, several thousand by now, who see the academy as top-heavy, dominated by an old-boy network in spite of the ongoing reforms, insufficiently concerned with ethical questions, and unprepared to develop new ways to support fundamental research. The union advocates the removal of all institutes from the jurisdiction of the academy, support for independent cooperatives, and the creation of private foundations with peer review. Union members believe these steps will free technology from its heritage of gigantomania and extensive development.[8]

Members of the union agree that scientists themselves should be given virtually unlimited freedom to decide who gets funding and at what level. This is more than a criticism of the Academy of Sciences, which, it is alleged, gave funding to bureaucrats and institute directors, not scientists. It also suggests a view of scientific activity as value neutral, with experts the only ones capable of deciding which projects to fund. This view seems to have wide support, as the creation of a new Committee of Social Expertise of the USSR indicates. The committee is a voluntary organization that intends to provide local and national government—in particular the Supreme Soviet—with "objective" scientific and technological advice.[9]

Some scholars believe that only through the resurrection of the social sciences and humanities can solutions be found for the economic, ecological, technological, and political problems facing the USSR. During the Brezhnev years the major force behind sociological approaches to Soviet socioeconomic problems was the academician Tat′iana Zaslavskaia. More recently, the academician Ivan Frolov has sought the creation of an Institute of Man (Institut cheloveka) to ensure the study of social institutions and human values during perestroika and, in particular, to consider the impact of new, "avant garde" technologies on society. A first step, dating from the fifties, was the creation of a general academic program, "Man, Science, Society: Interdisciplinary Research," which focused attention on social relations in socialism and on humanistic ideals, so as to go beyond nature (genetics and ecology) to consider nurture (ecology and environment). Frolov hopes to include the institutes of Sociology, Psychology, Philosophy, and History of Science and Technology in the program.[10]

While the creation of an Institute of Man suggests widespread belief in the social and cultural determinants of modern technology, most natural and life scientists and engineers continue to believe that scientific activity is value neutral. But a growing number of physicists and engineers are convinced of the need to turn away from an emphasis on big science and toward the humanities and social sciences. The president of the Academy of Sciences, G. I. Marchuk, considers the rise in anti-Semitism, the poor condition of social scientific research institutes, and the poor state of the social sciences in general to be "the consequence of long-reigning technocratism, of disregard for humanistic culture."[11]

The Communist Party, the Supreme Soviet, and Science and Technology

The importance of the Communist party also declined in science and technology policy with respect to newly formed groups of experts and the Supreme Soviet. At the national level, before the reforms instituted under Gorbachev, the central policy bodies—the academy, various state committees, the Council of Ministers, and GOSPLAN (the State Planning Commission)—resisted new approaches to funding, administration, and new fields of research. "Big science" received priority. Now, with the democratization of the administration and membership of institutes, along with decentralization of management and debates over the appropriate level of funding for fundamental and applied research and technology, all issues are fair game. Once the Party began encouraging the formation of independent groups of experts, its own role shifted into the spheres of housing and food, trade union activities, and public relations. The Party encouraged researchers to see it not merely as an expediter of matériel and equipment or an agent of ideological control.[12]

The Supreme Soviet has moved very slowly to fill the void left by the Communist party. In spite of great concern among many deputies about the place of modern technology in Soviet society, the Supreme Soviet has yet to focus on scientific issues. With the exception of the Greens, none of the new political parties have addressed technology in a systematic fashion.[13] This is not surprising for a number of reasons. First, nationality politics, economic uncertainties and debate concerning the future direction of the Soviet economy, and other pressing issues rightly have captured attention. Second, the Supreme Soviet lacks the expertise and staff to turn to an examination of science policy. While a series of standing committees and subcommittees—on science, education, and culture; on health; on informatics and transportation; on ecology—deal with related issues, the members of the committees are the overworked deputies themselves, who lack the staff to assist in deliberations.

The absence of expert staff advice or the creation of some sort of Supreme Soviet "office of technology assessment" also prevents measured consideration of science policy. Some deputies have called for the creation of a program like the congressional fellows program in the United States and an organization like the American Association for the Advancement of Science.[14] Finally, some deputies seem to reject the notion that the committees should be like previous Party bodies that "commanded." A member of the subcommittee on education and science, the astrophysicist V. L. Ginsburg, argues that it should merely bridge or arbitrate disputes in the solution of debates, assist in developing procedures to protect intellectual property, and more clearly define the role of the academy in terms of fundamental research and training, while leaving the State Committee for Science and Technology (GKNT) to be concerned with technology.[15]

The fact that many of the deputies of the Supreme Soviet are scientists and engineers will assist it in evaluating modern technology. There are 25 representatives of the medical profession, 22 natural scientists of whom 14 are physicists, and 22 social scientists in the 541-member councils of the union and nationalities—almost 11 percent —and a significantly larger number appear to have higher scientific or engineering degrees.[16]

To date, however, none of the legislative acts of the Supreme Soviet has touched specifically on environmental regulation or on technology, although it has passed a resolution on the poor state of the Chernobyl cleanup effort, and a series of laws currently under consideration are relevant to science and technology policy. One, the Law of the USSR on State Scientific-Technological Policy, is intended to establish general priorities and government responsibilities, including fixing the importance of fundamental research, and will address ecology, "informatization" (bringing the results of the computer and information revolutions to the USSR), energy policy, and the new role of independent scientific-technological expertise, especially concerning large-scale projects. A law on intellectual property has already passed.[17]

Initially, Gorbachev, too, manifested unbridled faith in scientists. He surrounded himself with an informal science advisory committee including such physicists as E. P. Velikhov and R. Z. Sagdeev and several economists. Owing to worldwide prestige earned by forty years of achievements, Soviet physicists gained significant authority in leading Party circles. The construction of the nearly completed 3,000-GeV proton accelerator, "UNK," and a new linear accelerator in Protvino, and the imminent completion of the T–15 tokamak fusion reactor at the Kurchatov Institute for Atomic Energy underscored the government's view that "big physics" should remain at the center of national attention because it benefits the economy, generates international prestige, and attracts international collaboration in a wide range of fields important to the Soviet economy. However, these advisers and their programs have become less prominent recently as the problems of nationality politics and the economic downturn have taken Gorbachev's attention. Gone also are the terms *"uskorenie"* (speeding up the assimilation of science and technology into the economy) and "scientific-technological progress" from Gorbachev's lexicon.

Nuclear Power, Space, and Big Technology

The Soviet nuclear power and space efforts show most clearly the extent to which technological arrogance dominated program decision making. As in the West, millions were spent on programs with limited social utility and questionable technical feasibility. These include a "project plowshares" to develop thermonuclear devices for geological engineering, a nuclear airplane like the U.S. ANP, and a series of satellites with nuclear-reactor power sources. But as a result of critical examination of R&D programs encouraged by perestroika and glasnost, physicists no longer command the public support they once did. Such significant failures as the Chernobyl reactor explosion and the loss of Phobos 1 and 2 contributed to the decline in confidence.

Physicists have maintained a firm conviction, however, that the development of large-scale technologies must continue, indeed is inevitable, and sense the need to circle wagons to defend their work. Whether in *Vestnik Akademii nauk SSSR, Energiia,* some other academy publications, or in public forums, physicists support extensive expenditures on nuclear, high-energy, and space research. While each of these areas now faces a skeptical, at times hostile, public reception owing to cost, potential environmental dangers, and modest spin-offs for the civilian sector, physicists continue to be technological determinists and find it inconceivable to halt technological advance. A. D. Sakharov, known in his last years more for his political conscience than for his physics, pushed for the construction of nuclear power stations underground as a way around safety concerns and public resistance.

The Chernobyl disaster revealed the extent to which technological arrogance long held sway among nuclear engineers. Soviet engineers and planners pursued an aggressive program of commercialization of nuclear energy through the prefabrication, if not mass production, of reactors and their components and such cost-cutting measures as less than adequate containment vessels. They believed that reactor technology is inherently safe, could be operated by nontechnical personnel, and that there is a technological fix for most problems.[18]

The Atommash plant on the Volga at Volgodonsk was the flagship of the atomic-energy industry and most visible symbol of technological arrogance. Intended to mass-produce eight VVER 1000 MWe reactor vessels and associated equipment annually by the early eighties, it has yet to deliver all eight. This is in spite of the fact that the massive plant actually consists of three production facilities, one of which is responsible for supporting the main plant with special tools and equipment and maintenance facilities, i.e., developing the tools to build the reactors. Atommash has been fraught with construction delays and a variety of shocking setbacks, including burst water mains and problems with utility lines. The foundation of the main foundry was built improperly, and it is reported that an entire wall collapsed. No more than half of the social services—movie theaters, workers' clubs, kindergartens, grocery stores—have been finished. Worse still, Volgodonsk is located in a region rife with industrial pollution, high cancer rates, and high infant mortality. As if to ignore what they have learned from the Chernobyl experience, the nuclear engineers who operate the Rostov nuclear power station near Volgodonsk (now coming on line) to this day reject public opposition as antiprogress. They hold that experts know best, especially in the face of such issues as the greenhouse effect.[19]

Until April 1986, the nuclear-power industry, like the space program, remained unassailable in the press and public. But glasnost permitted discussion of the unmitigated ecological disasters spawned by overzealous nuclear physicists at Chernobyl and Volgodonsk. Now they cannot generate public support. Even though there may be significant energy shortfalls as a result, at least ten nuclear power stations have been shut down; the Chernobyl type RBMK reactor has been abandoned; and the public has adopted a NIMBY (not-in-my-backyard) attitude toward nuclear power.

Physicists nevertheless question these attitudes. Researchers at such institutes as the Kurchatov Institute for Atomic Energy (KIAE)—the birthplace of Soviet nuclear engineering—the Institute of Theoretical and Experimental Physics (ITEF), and the Leningrad Institute of Nuclear Physics (LIAF) feel surrounded by a hostile public. Their research moves forward more slowly because of restrictions on the operation of their facilities. The experimental reactor at ITEF has been closed down since the Chernobyl disaster, and frustrated physicists cannot fathom why. At KIAE, nuclear facilities continue to operate, including the long-delayed T–15 tokamak, which commenced operation only this year. But the institute's weekly newspaper, *Sovetskii fizik,* reveals concern that Moscow's housing shortage continues to impinge upon the institute's once relatively isolated facilities and that public opposition will shut down its reactors, too.

Physicists at LIAF are particularly defensive about the safety of their reactor. In arguments reminiscent of those in the United States, they point out that there has never been an "event," that dosimeters around the site protect the public, and that LIAF employs a large number of local residents. At the beginning of the sixties, LIAF had the largest experimental reactor in the world. They cannot understand why they had to turn to Prime Minister Ryzhkov to gain long-delayed final approval on a new experimental reactor, the 100-megawatt "PIK," required to replace the VVR-M light-water reactor built in 1959 and now obsolete in spite of several upgrades in power, or why the local paper, *Gatchinskaia pravda,* did not support their efforts. After all, the site for LIAF was selected outside of Leningrad city limits precisely to enable the Leningrad Physico-Technical Institute to expand research in the areas of nuclear fission and high-energy physics.[20]

The formation of the Nuclear Society and the "Chernobyl Union," a national, voluntary, and independent social movement, created in 1989 to prevent future atomic accidents, to raise awareness of environmental issues, to organize radiation education, and to offer support to those who suffered in the Chernobyl disaster, has done little to dispel social concerns.[21]

The rejection of public opposition and the continued construction of the Atommash plant reflect the fact that Soviet engineers believe that they have achieved the level of technological sophistication necessary to mass-produce reactors. But Chernobyl and Atommash should have served as a lesson to go slowly in future programs and to include the public in the technology-assessment process. Even after the disaster, physicists continued to see technology as infallible—it was, after all, the operators of Chernobyl, not the plant itself, who were responsible for the disaster.

Supporters of nuclear power believe that the best way to deal with radiophobia is through education, with Gosatomenergonadzor providing the information.[22] Report cards on nuclear reactor "events" and "losses of power" and radiation levels in all cities, measured articles on the dangers of radioactivity, the ongoing activity in the Chernobyl "sarcophagus," and the cleanup effort appear regularly in the press. However, a recent study of nuclear power in Taiwan shows that giving the public more information about the industry actually increases perception of risk and makes the public less receptive to nuclear power, and in the Soviet case this is certainly the result.[23]

Nuclear energy receives more coverage in the Soviet press than any other technology.[24] A wide-ranging debate in the press among physicists, economists, and policymakers over the future of the industry has led to the development of the so-called "Chernobyl syndrome." The coverage includes reports on the absence of dosimeters for people who live in areas of Belorussia close to Chernobyl who have not been evacuated[25] and the sale of food tainted with cesium; exposés on the Chernobyl disaster, bureaucratic mismanagement of the cleanup and reclamation, the continued suffering of the Chernobyl victims, and confusion of dosages, rates, and exposure; and coverage of continued problems in the industry such as those in the United States at Hanford (Washington), Savannah (Georgia), and Fernald (Ohio), and in Semipalatinsk and Atomgrad on the Enisei,[26] including the public-relations fiasco surrounding the attempt to build a reactor in an active seismic region of the Tatar ASSR.[27] It hasn't helped that the republican press has published pictures of horribly deformed mutant animals (whose births predated Chernobyl)![28]

"Radiophobia" is rampant, encouraged by journalistic reports in such weeklies and journals of the writers's unions as *Literaturnaia Rossiia, Sibirskie ogni,* and *Novyi mir,*[29] much to the chagrin of scientists and policymakers alike. Poets condemn atomic energy as symbolic of the general lack of concern for the environment tied up in the creation of large-scale development projects.[30] Even children have been deeply affected by the tragedy of Chernobyl. Drawings of atomic power stations by French and Soviet schoolchildren reveal the extent of the problem. For the French child, the station sits in a pastoral setting, the sun shines overhead, and the workers, dressed in normal clothes, all smile. For the Soviet child, menacing, dark clouds hang over the atomic energy station, workers wear hard hats and stand sullenly at attention, and rockets with nuclear warheads ring the station.[31]

The entire atomic-energy industry and its research centers—Obninsk, Dubna, Pushchino, and others—no longer serve as magnets for talented Communist scholars.[32] But nuclear engineers—the *iadershchiki* —have held their own. They argue that economic development, fossil-fuel pollution (greenhouse effect, acid rain), and the inherent safety of atomic energy stations when operators are properly trained all require further investment in nuclear-power engineering. Their dreams of fusion-power reactors by the beginning of next century have survived. Their frustration with their loss of the prestige of the sixties and loss of innocence from the pre-Chernobyl days apparent, they still believe that modern science, although expensive, is the key to the future of Soviet society. Through glasnost and international contacts it will become "a powerful accelerator of progress." And if society and government do not wish to attract the best young talent to big physics with regular funding, if they require instead that fundamental research operate according to market mechanics, and if the best young talent then goes abroad, what will be left of the technological utopia?[33]

Space and the Frontier of Soviet Technology

Even space research, formerly the area of big science most important to the USSR for its display value, has fallen on hard times. In the fifties and sixties, no single area of Soviet scientific achievement captured as much of the national imagination as space: Sputnik, Iurii Gagarin, Luna, Voskhod, and a whole series of other "firsts" convinced many Soviet citizens that the radiant future might soon arrive.[34]

At one time, it was hard to find public enunciations of displeasure with any aspect of the space program. Now, however, criticism comes both from without and from within. R. Z. Sagdeev, initially a supporter of the space program who sought cooperation with the United States to help defray the costs of research, resigned from directorship of the Institute of Space Research (IKI) to return to his own research. He criticizes the pace and inadequate reach of perestroika. Sagdeev argues that only an open discussion of options and programs can resurrect public faith.[35]

The departure of Sagdeev from IKI slowed glasnost and democratization, according to one scientist. And what the space program precisely needs in a time of tight budgets is open discussion of competing programs to ensure selection of the best one and to avoid mistakes. But, Dr. V. Istomin claims, the top staff is still insulated from criticism from below, leading him to conclude that the authoritarian approaches of the past still hold force.[36] Scientists within the space community defend the Soviet programs from charges of high cost, insufficient social benefit, and excessive rate of failure with promises of economic "conversion"—benefits to the civilian sector.[37] Others, perhaps unaware of the significant problems facing the U.S. counterpart, call for the creation of a "NASA" with real glasnost concerning its financing to provide the coordination and long-term planning and to ensure real leadership—beyond that provided by the creators of rockets and satellites.[38]

However, at the urging of such newspapers as *Moscow News* and *Literaturnaia gazeta,* the whole space program has fallen into disrepute. Glasnost has proved to be a two-edged sword, with open discussion of past failures, present costs, and future fantasies leading to public mistrust. The failures of the Phobos probes, inefficient use of the highly touted Mir space station, technological backwardness especially concerning on-board computers, increasing frequency of launch delays and failures, and cost overruns during a time of economic uncertainly—all of these things help call into question the efficacy of space research. The Supreme Soviet has demanded greater control over space R&D and has cut the budget significantly. As in the United States, where space is no longer immune from criticism, manned missions in particular seem to be vulnerable.[39]

Leonard Nikishin, who sits on the *Moscow News* science desk, suggests the USSR should follow the example of the United States with significant budget cuts, since a country that advocated the Siberian river-diversion projects has a hard time saying "no" to big technology. "Wouldn't it be best at this stage," he asks, "to carry out a conversion of space research and limit our efforts to programmes that bring undoubted benefit: communication and navigation satellites, etc.?"[40] Oleg Moroz, the science editor of *Literaturnaia gazeta,* is even less supportive of the Soviet space program. He engaged the head of Glavkosmos, Aleksandr Dunaev, in a free-ranging discussion of the space program, singling out for criticism the budget and scope of the R&D; specific programs like Buran, Gorizont, Phobos; and the military aspects of the program.[41]

In other areas of big technology, the Brezhnev legacy of technological arrogance—such huge geological engineering projects as BAM (the Baikal-Amur Magistral), hydropower stations, canals, and the Siberian river-diversion project—faces close scrutiny. BAM undergoes constant ridicule, not the least for its long-overdue completion, although project directors and Party officials touted its alleged on-time completion. BAM was seen as a panacea to facilitate the exploitation of natural resources of the Soviet Far East. Today, the directors of the project admit the need to reexamine "Far East," the GOSPLAN program for development, to ensure commensurability with environmental concerns. But, like construction organizations put into place to build nuclear reactors, particle accelerators, and hydropower stations, Bamtransstroi, having gained institutional momentum based on personnel and equipment in place, turned to the construction of the 800-kilometer AIaM (Amuro-Iakutskaia Magistral). They hope to attract settlers, pointing to the creation of a series of small towns and villages that are "socially and culturally" complete. AIaM will be difficult to build in view of budget cuts, although access to Western firms through access to hard currency should help. But "Isn't it time for the Soviet Far East to be included in the process of *uskorenie?*" the director of Bamtransstroi asks.[42] In other words, engineers continue to see big technology as the key to "scientific-technological progress."

Unfortunately, development of fossil fuel and mineral resources in Siberia and the Far East has been conducted largely without attention to environmental and safety issues. As in the case of BAM, the explosion of the gas pipeline in Ufa raised significant issues about the interaction between cost, technology, and expert advice in the USSR when low-cost development remains a priority. The fact is that pipelines are built to withstand lower pressure than that at which they operate at larger diameters and are routed through inhabited areas to take advantage of existing train and automobile right-of-ways, all to save costs.[43] Local governments have begun to insist upon their inclusion in decision making concerning development projects, pushing for the preparation of environmental-impact statements.

It is clear, however, that none of the critics of modern technology recognizes that attitudes toward big technology as a panacea are not merely part of the Brezhnev legacy but are bound up in Soviet Marxist thought dating back to the first years of the revolution. In fact, growing belief in the power of modern computers, the market mechanism, and fundamental science to bring about perestroika indicate that this faith still holds in many circles.

The Computer as Panacea

Unlike space research, with its cost and all too visible failures, and atomic energy and its potential for significant environmental and human costs, the computer occupies a hallowed place in Soviet society. The problems of declining labor productivity, the need to accelerate information processing, CAD/CAM, and scientific research all seem at first glance to fall to the magic wand of the modern computer.[44] But for reasons that have been described elsewhere—the absence of a computer culture, fear of hackers and samizdat leading to strict institutional controls over hardware and software, lack of coordination in the computer industry between competing ministries and the Academy of Sciences, unreliable equipment, inability to produce computers, disks, and peripherals at anything near the level of demand—the USSR has entered the nineties without the computer. From schools to industry, while a significant improvement has occurred within the past five years, computers have not had the impact in the Soviet Union that one would expect in a developed country.

Under the leadership of the academicians D. M. Gvishiani, Andrei Ershov, and E. P. Velikhov, head of the new division of informatics of the academy, scholars are struggling to overcome the significant lag in computer science and technology. They have argued that through the "computerization" and "informatization" of Soviet society a whole series of intractable problems would give way. In the first years of the Gorbachev era their views remained unquestioned. They presented a series of proposals for "top-down" reforms to Party and government organs that they believed would lead to "computerization."[45]

The Communist party recognized the need for a new approach to computers—one that promoted broad social acceptance—only in the last years of Brezhnev. Taking note of the widening gap between computer technology in the USSR and the West, the Party turned to the leadership of the academy for a solution. The Politburo approved a fifteen-year national plan, promoted by Velikhov and announced in January 1985, to produce and introduce computer technology throughout industry and society. Its goals were to raise economic productivity and efficiency by accelerating scientific and technological progress (*uskorenie,* which has now disappeared from the Soviet lexicon). Regarding computers, this required an improved quality of equipment, an acceleration of production of more than 200 percent by 1990, and the introduction of new models of microcomputers and PCs. The plan did not come close to fulfilling the goals for 1990,[46] and Velikhov and Gvishiani are being blamed.

Velikhov and Gvishiani look to the United States and the broad application of computers throughout society, from the elementary school to the research institute, from the farm to the bank, and from the telephone to the hospital, as the example toward which the USSR must aspire. Their vision of "computerization," endorsed by the Division of Informatics, Computer Technology, and Automation of the Academy of Sciences and the State Committee for Computer Technology, embraced a top-down approach, where government resolutions will do the trick.[47] Gvishiani, director of the All-Union Scientific Research Institute of Systems Research of the Academy of Sciences, in findings presented to the Supreme Soviet suggests that no more than top-down reforms, millions of computers, and billions of rubles are needed.[48]

But others say that Gvishiani and Velikhov have failed to address the issue of social receptivity required of a true computer revolution through informatization. They suggest that Gvishiani recognized the storm his report would create inasmuch as deputies of the Supreme Soviet were told not to give the text of the project—a scant forty-page brochure—to journalists. Building computers and putting out manuals on their use are not enough. Central government expenditures on informatization average forty-eight billion rubles a year, but the results are invisible.[49] They argue that more modest steps like those taken by the Supreme Soviet to support intellectual property rights are needed, as are such initiatives from below as the formation of the Association for Artificial Intelligence.[50]

The attack on Velikhov has been less muted, in part because he has developed a reputation of being long on promises and short on accomplishments. How is it, critics ask, that in a country of one million physicists and billions of rubles of investment, there are no computers?

Once again, part of the answer is the absence of a computer culture and infrastructure. But Velikhov's inaction over five years is also inexplicable, and spending billions more is unpalatable.[51]

The failure has nothing to do with the absence of analysis of the place of computers and information in the USSR. In the eighties, a number of new visions for an information society were developed. Two of the major new assessments came from G. R. Gromov, a computer specialist whose name now graces the Institute of Cybernetics of the Ukrainian Academy, and the academician Ershov.[52] In a book entitled *National Information Resources,* Gromov documented the existence and importance of the information revolutions in the West, particularly in the United States, and the need to go beyond mainframes to PCs to trigger an information revolution in the USSR.

Ershov, who died in 1989, placed himself in the tradition of such futurologists as Alvin Toffler in coining the term "informatization." Informatization goes beyond "computer revolution." In what sounds much like the treatment given to information management by Daniel Bell in *The Coming of Post-Industrial Society* (1973), informatization is pulled by recent advances in electronics technology and is pushed by the rapid growth in the number of white-collar workers and burgeoning mass of information generated in modern society. Based on the belief that "knowledge and information are resources for which no substitutes exist," Ershov argues that "informatization" goes far beyond "automation." Automation pays little attention to the implications of computers for freedom of information, disestablishment of bureaucracy, and fundamental social change. Informatization, on the other hand, is consistent with early theories of the self-organizing and self-regulating systems of cybernetics, offers greater potential for individual creativity, and is more consistent with what the Soviet future seems to demand to enter the twenty-first century. Such publications as *Literaturnaia gazeta* have embraced Ershov's view. One of Ershov's leading supporters, the Leningrad scholar V. V. Aleksandrov, believes that only by placing value on information, by encouraging its transfer through decentralization of the economy, and by constructing telephone lines and cables and space relays is scientific-technological progress possible.[53]

If the goal is informatization, then the question becomes how ready is the USSR for this process? How are Soviet citizens exposed to computers? In the workplace, they are neither abundant nor unknown.

The problem here is the perpetuation of the myth of the mainframe as salvation for Soviet economic problems, rather than PCs to give greater exposure throughout society, although not surprisingly such industrial applications as numerically controlled machine tools and robotics also lag far behind the West. And in day-to-day life, who has seen a cashcard or credit card? The latter is a fascinating prospect: a Gosbank Visa card with the ability to acquire and transfer funds electronically. Steps have been taken in this direction, although officials acknowledge problems in developing the plastic magnetic information strip on the back of the Gosbank card, and I will remain skeptical until I can withdraw three hundred rubles from my account from a cashcard machine in the Kremlin Wall.[54]

The Market: Another Magic Bullet

A striking aspect of the change in attitude toward science and technology in the contemporary USSR is the growing belief that the U.S. system, once again, is a panacea for Soviet problems. From economic theory and the market mechanism, to the symbols of technology—in particular the computer, the automobile, and the telephone—most Soviet observers see *Amerikanizm* as a way out of the ongoing economic decay in the USSR. Some scholars have even called for a resurrection of the Taylorist Institute for the Scientific Organization of Labor, which found favor under Lenin, but whose staff and founder were arrested or eliminated under Stalin.[55] That a "workers' state" would embrace time-motion studies in view of the history of workers' responses to Taylorism is rather curious indeed.[56]

Among scientists and engineers, opinion about the value of the market is divided among those engaged in fundamental research and those in engineering and applied science.[57] Researchers in basic science see the market mechanism and *khozraschet* (economic accountability) as a threat to the financial security of their disciplines. Profit, cost-accounting measures, and individual gain at the expense of the *kollektiv* have no place in modern science, they argue. This sentiment is strongest in the institutes of the academy where scholars believe that the government must support research where the outcome or application cannot be predicted. (They assume, however, some social benefit at some point in the future.) They call for the creation of several different sources of funding, including grant and contract agencies modeled on the National Science Foundation that work according to peer-review principles.[58]

Engineers and technologists, on the other hand, embrace the market wholeheartedly as a means to accelerate scientific advances into the economy and a magnet for foreign investment. Several hundred scientific cooperatives have already formed based on market principles; their research foci include new ceramic superconducting materials, artificial intelligence, and biotechnology. Some scholars are critical of the cooperative movement in science. Since cooperatives are usually staffed by individuals from institutes who "free-lance" with state equipment, they argue that cooperatives are merely a modern way to steal state property and time. Most, however, see the cooperative as a way to raise tight funds—including hard currency—for research.[59]

Another organization born of perestroika in science and technology is the MNTK (*mezhduotraslevyi nauchno-tekhnicheskii kompleks*, or interbranch regional scientific-technical association). It was formed to overcome bureaucratic impediments to innovation in the USSR and to increase the productivity of research. This is especially the case in such "sunrise" industries as biotechnology, fiber optics, high-temperature superconductivity, robotics, and computers, where the USSR lags far behind the United States, Western Europe, and Japan. Each MNTK has an academy institute at its head, with engineering and design bureaus, experimental factories, and trade firms most likely from branch industrial ministries under its jurisdiction. Like the artificial intelligence and biotechnology firms that have arisen around universities throughout the United States, the MNTKs are supposed to help bridge the gap between scientific advances and the production process. A crucial issue is whether the associations will have the wherewithal and authority to succeed where their predecessors, the scientific-production associations (*nauchno-proizvodstevennoe ob''edinenie*, or NPO), failed. The difference between NPOs and MNTKs is that the former exist largely for solution of a specific technological problem in a specific branch or subbranch of industry, whereas MNTKs try to take an idea from science into production. Even supporters of the MNTKs expect them to encounter significant obstacles during a time of economic uncertainty in the face of parochial ministerial interests. The relative poverty of academy institutes and the absence of venture capital also present a problem.[60] But wherever engineers and scientists stand on the role of market mechanisms for science and technology, they all agree that for certain big technologies and military R&D the government must continue to be the major funding partner.

The play of market forces may be the most effective way of halting the momentum of large-scale engineering projects. The construction of nuclear reactors, particle accelerators, and the towns for the workers who build and operate them requires thousands of employees and millions of tons of equipment. In order to avoid unemployment and significant investment in transportation and other costs, Soviet planners have long sought to provide funding for projects that make use of workers and equipment in place. The linear accelerator for the Institute of Nuclear Physics in Novosibirsk (IIaF SO AN SSSR), for example, was approved for construction by the Council of Ministers at the site of the 3,000-GeV, "UNK," accelerator at Protvino near Serpukhov because of the presence of a huge construction brigade whose work on "UNK" was soon to be finished. The construction of the Rostov nuclear power station employed workers and equipment from the Atommash plant. The AI&M railway grew out of BAM. When workers no longer have internal passports and are free to move about according to market forces, the momentum that large engineering projects currently command, simply by virtue of equipment and manpower in place, will begin to wane.

Medical Science and Technology

Only in the areas of ecology and nuclear power have writers, officials, and the public focused more attention than in public health. The discussion includes all aspects of medicine and culture: the failure of the system to deliver what was promised in the past; problems of infant mortality and life expectancy; infectious diseases, cancer and heart disease, and the ecological sources of these problems; and the shocking backwardness in biomedical technology and its delivery. Epidemics of dysentery, hepatitis, and tuberculosis are now common in the USSR.

As in other areas of Soviet science and technology, a certain kind of technological arrogance long held sway in medicine. The manifestation of this attitude was a reliance on such quantitative indices of accomplishment as numbers of hospitals, beds, and doctors, and paradoxically, inadequate attention to modern medical techniques. Extensive regulations govern the numbers of days for treatment, and there are quotas on supplies and operations. The tendency is to try to fill beds, since unfilled beds might lead to a reduction of quotas. In this system medicines and technology have low priority, are of poor quality, and are ineffective.

The domination of the field by women led to the identification of the profession with low status, except in research, which is dominated by men. Medical care embraced "Fordist" notions, where dentistry, birthing, surgery, and abortion resembled assembly lines with six or eight patients to a room. Medical research also turned to the example of the U.S. system. Where else but in the USSR would a leading surgeon, S. N. Fedorov of the Moscow Research Institute of Eye Microsurgery, be so successful in promoting a surgical procedure for shortsightedness that is conducted on conveyor belts à la mass production?[61] And where else would surgeons be certified at the age of twenty-two after one year of specialized training to fit into the assembly line like a cog?

Glasnost and perestroika have changed the flavor of the debate over medical-care delivery and technology. Initially, most attention concerned the fact that the Soviet medical system served the elite, which had better access to modern technology than the vast majority of citizens. Next, as in other areas of Soviet science, exposés highlighted the domination of the medical research by individuals and their institutions, which delayed the introduction of new technologies seen as a threat to their control. This involved criticism of those seen as responsible for the creation of the administrative-command system of science and attacks on the huge scientific fiefdoms that accumulated during the Brezhnev period.

According to a series of articles that have appeared in weeklies, scandal sheets, and even *Vestnik Akademii nauk,* one such figure, Iu. A. Ovchinnikov, before his recent death a vice-president of the academy, director of the Shemiakin Biology Institute, main editor of a leading journal, member of the Central Committee, and deputy of the Supreme Soviet, apparently waylaid the development of an experimental blood substitute ("blue blood") produced by a provincial scientist without the resources of Ovchinnikov's institute; the scientist fell into disgrace and committed suicide. In another case, a *kandidat* of medical science developed a technique for using ultrasound to treat kidney stones. The medical establishment ignored and attempted to discredit the doctor, even after he received the endorsement of the chairman of the Council of Ministers, N. I. Ryzhkov.[62]

When faced with chapter and verse concerning accidents, age-specific death rates, infant mortality, and women's health issues, the citizen cannot help recognize that emphasis on numbers rather than delivery led to the poverty of Soviet health. Much has been made of the environmental sources of these significant problems, but there is growing awareness that technological backwardness, including the absence of such equipment as kidney dialysis machines, computerized axial tomography (CAT) scans, and standard emergency-room equipment, and such drugs as antibiotics, aspirin, and ibuprofen, is a major reason for high infant mortality, accident mortality, and rising age-specific death rates. It is increasingly acknowledged that investment as a percentage of GNP must grow substantially. AIDS is becoming a significant problem, made worse by the absence of basic technology—condoms and disposable needles.[63]

There are two problems that result from the widespread coverage of health problems, of production shortfalls, and of needless human suffering in the Soviet press, particularly in such journals of the writers' unions as *Raduga, Siberskie ogni, Iunost'*, and *Oktiabr'*. First, it is difficult to gauge the accuracy of some of the claims. Second, the reader becomes desensitized to the extent of the problem. Some general conclusions are possible, however. The 1972 resolution of the Council of Ministers "On Measures for Significant Improvement in the Production of Medical Instruments" notwithstanding, the Soviet medical-instrument industry meets no more than 50 to 75 percent of its production requirements, with promised factories never built.[64] According to the chairman of the Supreme Soviet committee on health, plans for production of medicines and medical equipment reach 50 to 60 percent of their targets, even for such elementary things as iodine, because of a backward pharmaceutical industry.[65] Low investment in medicine as a percent of GNP in comparison with other advanced countries has left the USSR twenty to thirty years behind the West in most medical technologies.

There appear to be two approaches to health and safety issues. One is a recognition of a need for some organizational reforms and increased investment, without substantially changing the face of Soviet medicine. Doctors lament Soviet health habits, high rates of smoking and drinking, poor housing and food, and lack of exercise; and they believe the public is as much to blame for the sad health of the average citizen as the medical profession. The widespread belief that universal free health care is a right, not a privilege, and a paradigm of the human body as a machine, which like a car often must be fixed, reinforce this view. Some even call for the creation of an organization like the American Medical Association to ensure preservation of a healthy moral climate and discussion of biomedical ethics on the highest plane possible.[66]

But most physicians seem to believe that a turn to *khozraschet* will solve most problems facing the medical profession, generating higher salaries and attracting talented researchers. They try to deflect criticism of such successful medical entrepreneurs as the eye surgeon S. N. Fedorov, his institute, and its orientation toward profit by pointing to the fact that narrow areas of specialization will be the exception. Besides, Fedorov earns *valiuta* (foreign currency) for Minzdrav, the Ministry of Health.

An increasing number of cooperatives, indeed, seem to provide better service. But, as has been documented in several U.S. cities, now doctors' offices and ambulance services wonder first about the ability of a patient to pay rather than his need for medical assistance. A homeopathic cooperative on Moscow's Highway of the Enthusiasts called "Lechenie i konsul'tatsiia" will not treat patients without proven ability to fork over thirty-five to forty rubles a day.[67] Some doctors also suggest that many cooperatives at this early stage of investment have less equipment than a majority of regional, average clinics.[68]

The media and government support the view that some radical changes are needed. *Moskovskaia pravda,* for example, established a new regular column, "Through the Eyes of the Patient." The column indicates that the public is usually satisfied with service, but most, especially the pensioners, find fault with the filth of the facilities and backwardness of the equipment. The Supreme Soviet also receives hundreds of letters and phone calls every day, but its committee on medicine answers few of them for lack of staff.[69]

In view of this sorry state of affairs, the Soviet public has responded to recognition of these significant problems with a growing fascination with holistic medicine, homeopathy, acupuncture (which had never been held in high esteem, but fell into disrepute after the Sino-Soviet split), and television psychotherapy.[70] This interest has found resonance in the spread of cooperatives in each of these fields and in the phenomenon of Anatolii Mikhailovich Kashpirovskii, whose television show features testimonials of complete recovery from various ailments rivaling those claimed by U.S. evangelists. Kashpirovskii's broadcast, once shown bimonthly, recently appeared as often as four times a week. While provoking the disdain of many in the medical establishment who see him as a quack and manipulator, he seems to have had as many devotees as any other Soviet TV show.[71] Another healer, Dzhuna Davitashvili, cures by the laying on of hands.[72] Some scholars call for official study of the extrasensory phenomena to verify—perhaps even license—such therapists as Kashpirovskii. One author, a representative of *Urdmurtskaia pravda,* asserts that certification is common practice in the West and that the Soviet public needs greater access to it.[73]

In spite of all this, there is growing attention to the way in which medical technology can improve the life of the Soviet citizen. Exhibitions of foreign medical equipment receive positive coverage in the daily press. An article in *Literaturnaia gazeta* raised the possibility of widespread use of artificial organs and skin by early next century. The article raised a series of ethical issues associated with the acquisition—even sale—of organs for transplant, decisions over who has access to the expensive procedures and technology, and the quality of life of such patients as Barney Clark after the implantation of artificial hearts. But in its final note, the article toasted the two-year anniversary of the second Soviet heart-transplant patient, as if to suggest that since the West does it, we ought to as well.[74]

The Environment, Biologists, and Ecologists

Biologists and environmentalists do not reject modern technology but call for research on a higher moral plane.[75] They question the unbridled faith of their physicist counterparts. They call for a technology-assessment process something along the lines of what goes on in the United States based on cost-benefit analysis (CBA). This reflects the belief that technology itself is neutral and what matters is whether it is abused or used properly, since, as Western critics of CBA argue, there is no way to include values and belief systems in calculations.[76]

Some scientists are aware that the technological-assessment process is often tilted in favor of development. This is because it is hard to derail projects once started, there is often a lack of public awareness of the process, and those who push technologies have the advantage of more complete information. The chairman of the Supreme Soviet committee on ecology, A. Iablokov, is concerned about what this means for Soviet "agrobusiness," which is dominated by the chemical industry. The result is an emphasis on poisonous pesticides rather than biological methods of control. The agrochemical industry anoints the Soviet land with more than two kilograms of pesticides per person per year. This situation has resulted, in part, from emphasis on such production indices as amount of investment, not additions to harvest or lowering of costs.[77] Soviet fascination with such modern technology as bovine growth hormone (bovine somatotropin, bST) is also evident. Firms such as Monsanto are beginning to explore the possibility of the sale of bST to the Soviets, at the same time that many states—Wisconsin, Minnesota, and Washington—have outlawed the sale of milk produced by cows treated with bST for a variety of socioeconomic as opposed to scientific reasons.

I have the impression that environmentalism runs stronger among Siberian scientists than those in Moscow or Leningrad, with the weekly *Nauka v Sibiri* publishing a series of government documents on the Siberian river-diversion project, the nuclear-energy industry, and the environmental degradation of Lake Baikal.[78] And while such journals as *Priroda* and *Energiia* tend to be less supportive of alternative-energy programs than nuclear power, even *Vestnik Akademii nauk SSSR* frequently publishes articles on ecology, alternative sources of energy, the design of ecologically more advanced automobiles, and so on. Even *Planovoe khoziaistvo,* a journal of economic planners, has added a section "Ecology and Economics" as a companion to "Scientific-Technological Progress."

The burgeoning ecology movement has begun to run into problems created by economic pressures. The government appears unlikely to sacrifice investment capital on scrubbers for coal-fired generators, water treatment plants, or land reclamation, when heavy industry and the consumer sector demand support. While scientists argue that mining, construction, and other industries must be subject to OSHA-like constraints to ensure worker safety, officials and planners resist expenditures in these areas. For example, asbestos, which after 1992 will be outlawed in the United States, figures prominently in three new all-union construction cooperatives formed in fall 1989. I saw workers in the center of Moscow systematically break up asbestos-coated pipes without respirators or special equipment of any sort. An uneducated public and a government more concerned with development are unlikely to adopt Western safety standards overnight.

On the other hand, local governments have shown increasing interest in ecological matters and now exercise growing veto power over development projects.[79] Only the Church enjoys the population's trust more than the Green movement in the USSR. Soviet citizens are more concerned about environmental pollution than any other problem, including food shortages, AIDS, ethnic tensions, alcoholism and others.[80] In fact, in 1990 there were more members of environmental groups in the USSR than of any other organization—including the Communist party.

The Writers' Unions and the Public: Soviet-Style Luddites?

The writers' unions play a critical role in exploiting and encouraging involvement in the ecology movement. From the Russian writers of the Siberian camp, who criticize the environmental damage wrought by hydroelectric power, chemicalization, and nuclear energy, to the writers of the union republics, who endeavor to show how development in the republics occurred at their expense in support of "Russia," many literary figures call for an abandonment of Soviet-style industrialization and a return to agriculture. Indeed, the entire January 1989 Plenum of the Union of Writers of the USSR was devoted to "Land, Ecology, and Perestroika." It criticized extensive development projects, the government, and the Academy of Sciences for their role. The writers nominated candidates for the elections to the Supreme Soviet whose positions represent what I call modern-day Luddism.[81] One successful candidate, Sergei Zalygin, editor of *Novyi mir* and founder and chairman of the association "Ecology and the World," went beyond the resolutions of the annual writers' union meeting, criticizing the Soviet system of resource development and the "global egoism" of the Soviet approach, calling for the adoption of measures amounting to "military eco-communism."[82]

What troubles the writers most of all is the fact that it has been extremely difficult to slow down the momentum of the "extensive methods" of the geological engineers—the canal builders and river diverters, the Baikal paper combines, the "professionals of gigantomania," as Zalygin calls them.[83] The problem is that big science has been co-opted by such government organizations as GOSPLAN,

Gosstroi, and GKNT, which fail to include ecological calculations in their work. As the Ukrainian writer Fedor Morgun argued, "We are not so rich that all of the insufficiencies of technology can be solved 'at the tail-end [*na khvoste*],' and therefore it is necessary to think how to avoid them already at the beginning of each project."[84] And failing to slow development and increase funding for enforcement and cleanup, the writers favor a moratorium until social and political control is established.[85]

Encouraged by the writers, the public has developed a growing mistrust of technology that balances scientists' newly found academic and political freedom. Judging by the press, the public holds scientists responsible for the failure of modern science and technology to improve their lives, while at the same time creating significant environmental and safety hazards. The public has become anti- and pseudo-scientific. Interest in holistic medicine, television psychotherapy, and mysticism is rampant. Many citizens cannot understand how it is possible to fund big science—reactors, satellites, and particle accelerators—when consumer goods, food, and medical care are in short supply. Thus, at the same time as the public shows interest in UFOs with sightings of extraterrestrials reported almost weekly,[86] the space program has lost support from its formerly most loyal ally—the Soviet citizen. This is symptomatic of the state of technology in the USSR today.

What Does the Future Hold?

In a way reminiscent of the United States in the seventies, Soviet society has begun to question the power of modern technology to improve the quality of life. In the United States the loss of faith in scientists, engineers, and their work was a result of the contrast between the successes of the Apollo project and putting a man on the moon and the failures of continued poverty and poor race relations in the cities. The Vietnam war, shown graphically on television, suggested that scientists produced weapons of horrifying technological accuracy but limited social utility. These concerns were coupled with the public's growing awareness of our environmental problems.

In the Soviet Union the criticism of technology may also be connected with Afghanistan, an expensive space program with few civilian spin-offs, and environmentalism. Certainly, glasnost and perestroika encouraged discussion of the appropriate relationship between technology and society. Glasnost revealed the extent of Soviet economic, political, and social problems, and perestroika encouraged examination of scientists, engineers, and their work, revealing how their research was approved, funded, and administered seemingly beyond social control.

The Gorbachev reforms resulted in a new relationship between experts, government, and the public. All scholars—physicists, biologists, engineers, and computer scientists—have organized into new associations both to defend their professional interests and to ensure their input in decision making. Local governments assert their prerogatives in science and technology policy, having grown concerned about the social and environmental costs of unregulated industries and nuclear power. And they have adopted a "NIMBY" attitude to any technology suspect in the least. The Supreme Soviet has also grown wary of big science, but owing to the pressures of economic and nationalities problems has yet to focus on science and technology. In this situation, industrial ministries, the Academy of Sciences, and various state committees tend to be the major purveyors of technology, captivated by the momentum of past projects and legions of workers and equipment; scientists and engineers wish to be the final arbiters in choosing among competing projects; and the public, supported by the writers' unions and many social scientists, has grown skeptical of any development project. Having informed the public about nuclear, space, and medical research, scientists succeeded only in creating more opposition to their programs. Yet the danger of the underestimation of the unanticipated social costs of modern technology remains, owing to the continued faith of most scientists and engineers in science as a panacea.

Herbert Marcuse identified the dilemma facing Soviet society in his discussion of technological rationality in *One-Dimensional Man*. Technology can be used to create the illusion of freedom, of plenty, of choice and yet can act to contain change. It can also be used to help achieve the goal of modern societies: to alleviate man's struggle for existence. The latter view is supported by Soviet scientists and engineers.[87] Big technology, once a panacea supported by ideological precepts embedded in Soviet Marxism, faces an uncertain future. Owing to a view of technology as value neutral, other technologies have supplanted the reactor as the key to the Soviet future: the market mechanism and the access to foreign currency it brings, the computer, the

U.S. system of the organization of science, and so on. This view ignores the fact that with market mechanisms firmly in place, decisions will be made that reflect some measure of efficiency rather than values or beliefs, obscuring the question of which technologies *ought* to be supported. Throwing money at the computer industry will not necessarily create the conditions needed for broad social receptivity. And giving scientists and engineers increasing latitude to make decisions about their work without social and political controls will not necessarily ensure that they make the right decisions.

Notes

1. G. A. Cohen, *Karl Marx's Theory of History: A Defence* (Oxford: Oxford University Press, 1978).

2. Donald MacKenzie, "Marx and the Machine," *Technology and Culture*, no. 3 (1984): pp. 473–502.

3. See, for example, Nikolai Bukharin, *Historical Materialism* (Ann Arbor: University of Michigan Press, 1969), p. 124.

4. Hans Rogger, "*Amerikanizm* and the Development of Russia," *Comparative Studies in Society and History,* vol. 23, no. 3 (July 1981): pp. 382–420; and Kendall E. Bailes, "The American Connection: Ideology and the Transfer of American Technology to the Soviet Union, 1917–1941," ibid., pp. 421–48. *Odnoetazhnaia Amerika* (One-story America), by Il'ia Il'f and Evgenii Petrov, was published in the journal *Znamia* (1936), nos. 10–11) and subsequently as a separate book (Moscow: Goslitizdat, 1937).

5. As Parrott points out, even after the decision to pursue economic autarky in the USSR under Stalin, a debate persisted over the extent to which the USSR ought to embrace Western science and technology—a debate that was not resolved in favor of a decision to enter fully the world economic system until the Gorbachev revolution. See Bruce Parrott, *Politics and Technology in the Soviet Union* (Cambridge: MIT Press, 1983).

6. For discussion of contemporary Soviet science policy, several points of which are touched upon here, see Paul Josephson, "Scientists, the Public and the Party under Gorbachev," *Harriman Institute Forum,* vol. 3, no. 5 (May 1990).

7. Among the first societies created in the Gorbachev period is the Physics Society, formally established in November 1989 although meeting periodically for two years beforehand. It is the successor of the Russian Association of Physicists, founded in 1919, which disappeared in 1931. Besides engineers, physicists —at 400,000 strong—are the largest group of academic specialists in the USSR. See S. P. Kapitsa, "Doklad na uchreditel'nom s"ezde fizicheskogo obshchestva SSSR," 17 November 1989, Moscow, mimeographed. See also Kapitsa, "Vozrozhdennoe obshchestvo," *Priroda,* 1990, no. 3 (1990): pp. 71–77.

8. Interview with M. D. Frank-Kamenetskii, Moscow, 29 March 1990; and V. Chernikov, "Ostanovim li my utechku mozgov," *Sovershenno sekretno,* no. 1 (1990): pp. 24–25.

9. N. Krivomazov, "Parallel'nye—skhodiatsia," *Pravda* (12 November 1989): p. 1. Yet another manifestation of concern for the changing role of science and technology in the USSR has been a call for the creation of a Russian Academy of Sciences. Those most active in this movement are more conservative than those gravitating to the Union of Scholars. Primary among them are academicians from Moscow, Leningrad, and in particular Akademgorodok and the Siberian branch of the academy. Other supporters include Sergei Zalygin, N. Dubinin, and Valentin Rasputin. They believe that "Russian" science has served all republics through the USSR academy at the expense of the Russian republic. The debate about the Russian academy has now been the subject of a discussion within the Presidium of the USSR academy, with a vocal minority calling for the creation of a new Russian Academy of Sciences (see F. Borodkin, "Zachem nuzhny nadstroiki," *Nauka v sibiri,* no. 38 [29 September 1989]: p. 2; N. Dubinin et al., "U poroga tysiacheletiia," *Sovetskaia Rossiia* [12 November 1989]: p. 3; "Kakoi byt' Rossiiskoi akademii," ibid. [24 November 1989]: p. 3; V. Denisov, "Kontury akademii," *Sovetskaia Rossiia* [3 December 1989]: p. 2; A. Lepikhov, "Byt' li Rossiiskoi akademii?" *NTR,* no. 21 [1989]: pp. 1, 6; L. Usacheva, "Zaglianem v budushchee," *Poisk,* no. 28 [9–15 November 1989]: pp. 1, 3; and "Kakoi byt' Rossiiskoi akademii nauk," *VAN,* no. 2 [1990]: pp. 48–78).

10. I. Frolov, "Vozvrashchenie k cheloveku," *Poisk,* no. 16 (August 1989): pp. 1–2. See also K. Smirnov, "Perevernut' piramidu," *Izvestiia* (15 May 1989): p. 2, which discusses the results of a conference in Moscow called to discuss the creation of a "Center of Science of Man." Frolov's interdisciplinary approach has drawn praise from a number of scholars. See, for example, N. N. Moiseev, "Chelovek vo vselennoi i na zemle," *Voprosy filosofii,* no. 6 (1990): pp. 32–45.

11. G. I. Marchuk, "Perestroika fundamental'nykh issledovanii: tseli, zadachi, perspektivy," *Vestnik Akademii nauk SSSR,* no. 5 (1990): p. 42.

12. Beginning with a meeting of Party members of academy institutes in Zvenigorod in late November 1989, a series of national gatherings of party scientists and officials has been held to discuss the social and political context of modern science; the role, status, and prestige of the scholar in conditions of socialism; and science and technology policy (interview with V. I. Safarov and E. A. Tropp, senior physicists, Leningrad Physico-Technical Institute, 19 December 1989, Leningrad; V. Abramov, "Ser'eznyi rubezh," *Sovetskaia Rossiia* [20 November 1989]: p. 3; Lidiia Usacheva, "Kakoi ia khochu videt' partiyu?" *Poisk,* no. 32 [7–13 December 1989]: pp. 1, 3; and Dr. Liudmila Vartazarova, "Predlozhenie k proektu ustava KPSS" [mimeographed]).

13. This sense is based on a sampling of such newspapers as *Demokraticheskaia Rossiia* (paper of the Democratic party of Russia), *Uchreditel'noe sobranie* (the Democratic Union), *Soglasie* (the Lithuanian Movement for Perestroika), and *Pozitsiia* (the Social Fund 'Sodruzhestvo').

14. Kim Smirnov, "Chemu uchit' sia deputatu," *Izvestiia,* (5 December 1989): p. 4.

15. "Vyrastit' novyi sad," *Poisk,* no. 22 (September 1989): pp. 1–2.

16. *Izvestiia,* 1 June 1989, pp. 2–3.

17. G. I. Marchuk, *Vestnik Akademii nuak SSSR,* p. 36.

18. See Paul Josephson, "The Historical Roots of the Chernobyl Crisis," *Soviet Union,* vol. 13, no. 3 (1986): pp. 275–99.

19. Ivan Kunitsyn and Aleksei Nikolaev, "Po Donu guliaet ... 'mirnyi'

atom," *Iunost'*, no. 4 (1990): pp. 40–45.

20. Interviews with academician M. A. Markov, 29 November 1989, Moscow, and A. I. Il'in, secretary of Party committee, Leningrad Institute of Nuclear Physics, 18 December 1989, Leningrad.

21. On the Nuclear Society, see *Sovetskii fizik*, the weekly publication of the KIAE. On the Chernobyl Union, see *Ustav soyuza Chernobyl'*, 27 June 1989, and *Chernobyl': sobytiia i uroki* (Moscow: Izdatpolit, 1989), a handbook that was put out with the help of the Chernobyl Union. When in the USSR in the fall of 1989, I saw an announcement of the formation of the Chernobyl Union in *Moscow News* (1989, no. 46, p. 2), wrote the post office box number supplied, met with its founders, and received copies of its literature.

22. A. Illesh, "AES bez sekretov," *Izvestiia* (5 May 1989): p. 2.

23. "Fear and Loathing of Nuclear Power," *Science*, vol. 256 (5 October 1990): p. 28.

24. I have not used *Letopis' gazetnykh statei* or *Letopis' zhurnalnykh statei* to provide a basis for this assertion.

25. Sergei Trusevich, "Zapretnaia zona dlia dozimetrov?" *Poisk*, no. 22 (September 1989): p. 3, and Vasil' Iakovenko, "Zemlia atomnogo vspolokha: Belorussiia," *Literaturnaia Rossiia*, no. 43 (27 October 1989): pp. 6–7.

26. According to M. Iur'ev, "Rentgeny Eniseia," *Nauka v sibiri*, no. 40 (13 October 1989): pp. 4–5, swimming and fishing are strictly prohibited on the Enisee since dosimetry shows that the level of gamma radiation activity of the water exceeds background by six to eight times and greater. Iur'ev is surprised that two-headed fish have yet to be found.

27. Iu. Kotov, "Grozit li bezrabotitsa stroiteliam AES?" *Sovetskaia kul'tura*, (23 November 1989): p. 3. For a study of the construction of a nuclear power station in the United States on an earthquake fault, and the interaction of engineers, utilities, regulators, and scientists, see Richard L. Meehan, *The Atom and the Fault* (Cambridge: MIT Press, 1984).

28. "Mutantov—dvoe. A prichin?" *NTR*, no. 16 (1989): p. 7.

29. Grigorii Medvedev, "Chernoybl'skaia tetrad'," *Novyi mir*, no. 6 (1989): pp. 3–108; and Iurii Shcherbak, *Chernobyl: A Documentary Story*, trans. Ian Press, forward by David R. Maples (New York: St. Martin's Press, 1989). (Originally published in *Iunost'* in 1987 and in the Ukrainian journal *Vitchyzna* in 1988.)

30. Viktor Goncharov, "A-ES," *Literaturnaia Rossiia*, no. 43 (27 October 1989): p. 7.

31. V. Pokrovskii, "Strasti vokrug AES," *NTR*, no. 7 (1989): p. 1.

32. Obninsk, site of one of the world's first nuclear power reactors to generate electricity, a city where excited young scientists pursued visions of the construction of communism based on an unlimited supply of nuclear energy, was in 1990 a city without the potential for discoveries, dominated by Party bureaucrats and scientists intolerant of dissent, and relegated to consulting work for the milk industry (Nonna Chernykh, "Pochemu ugasaet nauka v gorode nauki?" *Ogonek*, no. 4 (1990): pp. 9–12.

33. Vanda Beletskaia, " 'ITER'—2003. Prikonchit li khozraschet nauku?" *Ogonek*, no. 33 (1990): pp. 1–2. This article is based on an interview with the academician B. B. Kadomtsev, head of the KIAE T–15 tokamak project.

34. On the history of the Soviet and U.S. space programs, see Walter McDougall, *The Heavens and the Earth* (New York: Basic Books, 1985); John Logsdon, *The Decision to Go to the Moon* (Chicago: University of Chicago Press, 1976); and Thomas Canby, "Are the Soviets Ahead in Space?" *National Geographic* (October 1986): pp. 420–59. On the culture of space and public attitudes in the early years of the Soviet program, see Paul Josephson, "Rockets, Reactors and Soviet Culture," in Loren Graham, ed., *Science and the Soviet Social Order* (Cambridge: Harvard University Press, 1990), pp. 168–91. On the current debates in the United States over space policy and criticism of mismanagement at NASA, see H. Guyford Stever and David L. Bodde, "Space Policy: Deciding Where to Go," *Issues in Science and Technology* (Spring 1989): pp. 66–71; John M. Logsdon, "The Space Shuttle Program: A Policy Failure," *Science,* vol. 232 (May 1986): pp. 1099–1105; Sally K. Ride, *Leadership and America's Future in Space,* A Report to the Administrator, NASA, Washington, August 1987; and Robert Bless, "Space Science: What's Wrong at NASA," *Issues in Science and Technology* (Winter 1988–89): pp. 67–73.

35. Roald Z. Sagdeev, "Science and Perestroika: A Long Way to Go," *Issues in Science and Technology,* vol. 4, no. 4 (Summer 1988): pp. 48–52; and Roald Z. Sagdeev, "Soviet Space Science," *Physics Today,* vol. 41, no. 5 (May 1988): pp. 30–38. See also A. Pokrovskii, "Kuda letim?" *Pravda* (15 November 1989): p. 3.

36. V. Istomin, "Ukhaby na puti k zvezdam, ili kosmicheskaia nauka v poru glasnosti," *Literaturnaia gazeta* (26 April 1989): p. 11. This article shares the same page with articles reprinted from 1967 by the German physicist Max Born and the academic A. A. Blagonravov, who debated the utility of space research, with Born finding the expenses too great and oriented toward the military.

37. N. Krivomazov and Iu. Khots, " 'Granit' ne kamen': ne drobit' " *Pravda* (13 November 1989): pp. 1, 3.

38. B. Konovalov, "Kosmonavtika na rasput'e (beseda s V. A. Shatlovym)," *Izvestiia* (17 May 1989): p. 2.

39. See, in particular, Boris Olesyuk, "Space Exploration—A Wasteful and Inefficient Effort," *Moscow News,* no. 30 (1990): p. 15. For a defense of the MIR space station, see the statement by three cosmonauts, A. Volkov, S. Krikalev, and G. Nechitailo, "Ot semian do sadov," *Pravda* (22 October 1989): p. 3.

40. Leonard Nikishin, "Editor's Note," *Moscow News,* no. 30 (1990): p. 15.

41. Oleg Moroz, "Ekonomit' li na kosmose?" *Literaturnaia gazeta* (20 December 1989): p. 11. Dunaev did his best to parry criticism by calling attention to past accomplishments, in particular the achievement of military parity with the United States in spite of the relatively low level of support in the USSR.

42. E. Basin (deputy minister of Transporation Construction, director of Bamtransstroi), "BAM: Skoro pusk," *Pravda* (1 November 1989): p. 2.

43. S. Denisov, "Pust' grom ne grianet," *Pravda* (11 October 1989): p. 2.

44. For a discussion of early attitudes toward the computer in the United States, see Paul Ceruzzi, "An Unforseen Revolution: Computers and Expectations, 1935–1985," in Joseph Corn, ed., *Imagining Tomorrow* (Cambridge, England: Cambridge University Press, 1986), pp. 188–201.

45. For a thorough discussion of the recent history of computerization in the USSR, see Richard W. Judy and Virginia L. Clough, "Implications of the Information Revolution for Soviet Society: A Preliminary Inquiry," Hudson Institute

Report, HI–4091-P, Indianapolis, 9 January 1989. See also S. E. Goodman and W. K. McHenry, "Computing in the USSR: Recent Progress and Policies," *Soviet Economy*, vol. 2, no. 4 (1986): pp. 327–54; and Loren Graham, "Science and Computers in Soviet Society," in Erik Hoffmann, ed., *The Soviet Union in the 1980s* (New York: Academy of Political Science, 1984), pp. 124–34.

46. Regarding failure to reach the schools, see M. Basina and T. Kamchatova, "Tekhnika na grani . . . primitiva," *Leninskie iskry* (Organ of the Leningrad Obkom and Gorkom VLKSM) (2 December 1989): p. 1.

47. "Novye grani davnei problemy," *NTR*, no. 8 (1989): p. 3.

48. V. Abramov, "Kontseptsiia est'—komp'iuterov net," *Sovetskaia Rossiia* (20 March 1990): p. 2.

49. Sergei Panasenko, "Mirazhi informatizatsii," *Sotsialisticheskaia industriia* (19 December 1989): p. 2.

50. D. Pospelov, "Iskusstvennyi intellekt: nashi bedy i trudnosti," *NTR*, no. 7 (1989): p. 7.

51. N. Loginova, "Vzryvoopasnaia tishina," *Literaturnaia gazeta* (13 December 1989): p. 11.

52. Much of this discussion is based on Judy and Clough, *op cit.*

53. S. Samoilis, "K informatsionnomu obshchestvu," *Leningradskaia pravda* (22 March 1990): p. 2.

54. V. Gorchakov, "Elektronnyi rubl'," *Komsomol'skaia pravda* (13 May 1989): p. 4.

55. See A. G. Lopatin, "Vozrodit' NOT," *EKO*, no. 1 (1990): pp. 138–43. On the early history of Taylorism in the USSR and Europe, see Charles Maier, "Between Taylorism and Technocracy: European Ideologies and the Vision of Industrial Productivity in the 1920s," *Journal of Contemporary History*, vol. 5 (1970): pp. 27–61, and Kendall Bailes, "Aleksei Gastev and the Soviet Controversy over Taylorism, 1918–1924," *Soviet Studies*, vol. 39, no. 3 (July 1977): pp. 373–94.

56. On the reception of Taylorism, see Hugh G. J. Aitken, *Scientific Management in Action: Taylorism at Watertown Arsenal, 1908–1915* (Princeton: Princeton University Press, 1985); and Samuel Haber, *Efficiency and Uplift: Scientific Management in the Progressive Era* (Chicago: University of Chicago Press, 1964).

57. This brief discussion is based on a reading of articles in *NTR*, *Poisk*, *Nauka v Sibiri*, *Vestnik Akademii nauk SSSR*, *Kommunist*, *EKO*, and a few other journals.

58. The physicist A. N. Skrinskii, academic secretary of the division of nuclear physics of the academy, worries that the most difficult and serious task regarding scientific-technological progress is "to attract the attention of society to the necessity of serious and active support of the development of fundamental science as the basis of technological progress and the most important component of all culture" (A. N. Skrinskii, "Traditsii i obnovlenie," *Energiia-impul's* [newspaper of the Institute of Nuclear Physics of the Siberian Academy of Sciences], no. 1 [April 1990]: p. 1).

59. V. Dadykova, "Kooperativy v nauke: zlo ili blago," *Nauka v Sibiri*, no. 6 (17 February 1989): p. 2.

60. A. Antipov, "MNTK—Deti perestroiki," *Kommunist*, no. 15 (1989): pp.

86–88. On NPOs, see Julian Cooper, "Innovation for Innovation in Soviet Industry," in Ronald Amann and Julian Cooper, eds., *Industrial Innovation in the USSR*
(New Haven and London: Yale University Press, 1982), pp. 456–70. Some 150
NPOs were created by 1978. The hope was that the NPO would accelerate the
speed with which scientific advances found their way into production. But since
GOSPLAN treated them as separate organizations they often became mechanical
conglomerations whose activities were poorly coordinated.

61. The operation "block" of this institute, which consists of nine operation
rooms, is equipped with eighteen surgical microscopes and tables where up to 150
operations are performed each day. "The idea of automating the prevailing eye
surgery interventions . . . with separation of each operation into five equally time-
consumed stages . . . was set in motion in June 1984"(*Intersectoral Research and
Technology Complex "Eye Microsurgery"* [Moscow: RSFSR Health Ministry, n.
d.], p. 29).

62. Vladislav Starchevskii, "Kogda zhe pridet vremia ubirat' kamni?"
Nedelia, 1989, no. 40 (1540) (1989): p. 7.

63. To date several hundred school children have been infected by contami-
nated needles. Growing coverage of AIDS ("*SPID*") in the press reveals the
absence of a coordinated national policy. Even "*SPID-INFO*" newspaper, a collection of translations from Western sources whose applicability to the Soviet
situation might be questioned, appears irregularly and in inadequate numbers.

64. V. Matukhnov and E. Zelenov, "So rzhavym skal'pelem," *Pravda* (18
October 1990): p. 3.

65. Elena Dikun, "Ne bolet! Vygodno kazhdomu," *Nedelia*, no. 52 (1989):
p. 7. See also, I. Krasnopol'skaia, " 'Izvinite, lekarstv net . . .' " *Moskovskaia
pravda* (22 December 1989): p. 3.

66. N. Kharitonova, "Prizrak chistogana," *Sovetskaia Rossiia* (27 October
1989): p. 4.

67. L. Ivchenko, "Grimasy 'LIKa'," *Pravda* (30 September 1989): p. 3.

68. E. Andzelevich, "Ne nash bol'noi," *Izvestiia* (4 October 1989): p. 3.

69. Elena Dikun, *Nedelia*, p. 7.

70. "Zdorov'e na pul'se," *Vecherniaia Moskva* (25 November 1989): p. 2; and
N. Krivomazov, "Ruku na pul's," *Pravda* (23 October 1990): p. 8.

71. "Fenomen Kashpirovskogo," *Sovetskaia kul'tura* (28 November 1989):
p. 6. The cult of Kashpirovskii continues to be spread in A. Morgovskii, *Seansy
A. Kashpirovskogo. Zagadki, legendy, real'nost'* (Moscow: Prometei, 1990), a
positive, but usually fair treatment. A discussion at the Institute of Philosophy in
November 1989 indicated some support for Kashpirovskii among professional
psychologists. See N. Kharitonova, "Glaza v glaza," *Sovetskaia Rossiia* (26 November 1989): p. 6.

72. Elena Agapova, "Novoe zvanie Dzhuny: rytsar'-komandor," *Krasnaia
zvezda* (5 November 1989): p. 4.

73. O. Evseeva and A. Valentinov, "Kto pomozhet . . . ekstrasensu?"
Sotsialisticheskaia industriia (25 October 1989): pp. 3–4.

74. Leonid Zagal'skii, "Chelovek iz zapchastei," *Literaturnaia gazeta* (1 February 1989): p. 11.

75. For a recent discussion of environmentalism in the USSR, see Robert G.
Darst, Jr., "Environmentalism in the USSR: The Opposition to the River Diver-

sion Projects," *Soviet Economy*, vol. 4, no. 3 (1988): pp. 233–42; and Philip Micklin and Andrew Bond, "Reflections on Environmentalism and the River Diversion Projects," ibid., pp. 253–74.

76. For a criticism of the quantitative methodologies of technology assessment for excluding social impact and implications, see Ida Hoos, "Society Aspects of Technology Assessment," *Technological Forecasting and Social Change*, vol. 13 (1979): pp. 191–202.

77. A. Iablokov, " 'Esli khotim vyzhit'... '," *NTR*, no. 15 (1989): pp. 4–5.

78. See, for example, V. Ermikov, N. Meshkova, and N. Pritvits, "Bor'ba za Baikal," *Nauka v Sibiri*, no. 8 (3 March 1989): pp. 2–3.

79. Such newspapers as *Vecherniaia Moskva* now publish an "Ecological Diary" on the work of the City Committee on Environmental Preservation.

80. V. Lupandin and G. Denisovsky, "The 'Greens' Coming to the Fore," *Moscow News*, no. 22 (1990): p. 7.

81. "Zemlia, ekologiia, perestroika," *Literaturnaia gazeta* (18 January 1989): pp. 1–4.

82. Sergei Zalygin, "Gosudarstvo i ekologiia," *Pravda* (23 October 1989): p. 4.

83. Zalygin, "Professionaly ot gigantomanii," *Literaturnaia gazeta* (8 February 1989): p. 11. See also L. Filipchenko, "Baikal'skii sindrom," *Izvestiia* (4 May 1989): p. 2; and "Aral–88," *Literaturnaia gazeta* (19 April 1989): p. 12.

84. Fedor Morgun, "Ekologiia: 12-u chas," *Raduga*, no. 12 (1989): p. 114.

85. The flavor of the far right, represented by Pamiat' and Otechestvo, is stranger still. The program of Pamiat', as published in *Literaturnyi Irkutsk* and other Siberian papers, argues that even science and technology have fallen prey to a "Judeo-masonic" conspiracy that operates on the principle "to free science from the Russians" and the overattestation of Jews in research institutes.

86. Lev Aksyonov and Boris Zverev, "Aliens Visit Voronezh: Eyewitness Accounts," *Moscow News*, no. 43 (1989): p. 7.

87. Herbert Marcuse, *One-Dimensional Man* (Boston: Beacon Press, 1964).

Innovation Strategies in Centrally Planned and Transition Economies

Susan J. Linz

It has become increasingly apparent that the institutional, organizational, and motivational structures in the Soviet economy are changing. Indeed, given the reception of the academician Stanislav Shatalin's 500-day program for the transition to a market economy,[1] it is more probable now than ever before that the outcome of the current reform program will be a market-oriented economy. If the transition proceeds along the lines proposed by Shatalin, central authorities no longer will dictate the composition of current production or the mix between current and future production.

If the transition succeeds in creating a decentralized decision-making environment where prices move freely in response to supply and demand conditions, profitability considerations will require Soviet enterprise managers to reevaluate their production strategies. During the transition period, firms must respond to constraints imposed as a consequence of dismantling the centralized allocation mechanisms when making current production decisions. Decisions regarding future production, however, are unlikely to be significantly affected by the transition period. Firms that have no confidence in the successful transition of the Soviet economy to a market economy will be unlikely to

This paper was prepared for presentation at the American Association for the Advancement of Slavic Studies meetings in Washington, D.C. (October 1990) as well as at the Ohio State University conference on "Technology, Culture, and Development: The Experience of the Soviet Model" (October 1990). The author wishes to thank Paul Segerstrom, Bruce Allen, Walter Adams, Holland Hunter, Seymour Goodman, and Loren Graham for their comments on an earlier version of the paper.

commit resources to future production. Rather, these firms will use existing technology to satisfy current demand. This paper ignores the pessimistic firms.

Firms that decide to invest in the development or diffusion of new technology, to invest in projects that may take years or decades to complete, reveal their expectations that the transition to a market economy will be successful. Clearly, the nature and pace of the transition will affect their expectations of future market conditions and thus influence the level or rate of innovative activities. But, given their expectations, these managers will necessarily be guided in their innovation decision making by the same factors that guide the innovation decision making of managers in market economies: invest in the development or diffusion of new technology whenever the present discounted value of expected profits is greater than the cost of the expenditures involved, that is, whenever anticipated reward exceeds anticipated risk.

The goal of this paper is to identify the optimal managerial strategy with respect to the development and diffusion of new technology in light of the changing economic environment in the USSR. Like firms in market economies, Soviet firms will have to choose between innovating (investing in the research and development of new products or processes) and imitating (investing in existing products or technologies). Because innovation and imitation activities are costly both in terms of physical and financial resources, government policies that raise or lower the cost of innovation are likely to influence R&D strategies. It is assumed that Soviet government policies regarding innovation will be dominated by the desire to close the technological gap with the West without jeopardizing the availability of consumer goods. Thus, leaders are likely to adopt policies during the transition period that unambiguously promote economic growth, even if these policies result in a slower rate of technological advance than would be possible given existing resources.

The chapter is divided into three sections. The first section overviews the current innovation environment in the Soviet Union, using Soviet and Western literature and interviews with recent Soviet emigrants to the United States as sources of information.[2] The second evaluates the potential impact on the Soviet innovation environment of (1) Shatalin's 500-day plan for the transition to a market economy announced in September 1990 and (2) the alternative, more conservative, proposal offered by Mikhail Gorbachev in October 1990. Section three analyzes managerial strategies for innovation and evaluates alternative policies to promote the development and diffusion of new technology.

Soviet Innovation Environment

The Soviet innovation environment is characterized by a discrepancy between the relatively high level of expenditures on R&D and the relatively low rate of developing and diffusing new technology. Bergson (1983) and Thornton (1986) document expenditure and employment patterns that reflect the Soviets spending over time an increasing share of national income on basic and applied research in the USSR. In 1970, for example, national income totaled 289.9 billion rubles (current prices), with 4 percent going to expenditures on science; by 1985 national income had risen to 567.9 billion rubles (current prices), of which expenditures on science totaled 5 percent. During this same period (1970–85), the total number of scientific workers in the USSR rose from roughly 927,700 to 1,400,000 (Thornton 1986, table 6).

Despite the high priority accorded to technological advance, the Soviet Union has been unable to achieve planned gains.[3] Soviet and Western researchers have analyzed at length the innovation process in the USSR in an effort to understand why the Soviet Union appears to lag behind the West at each stage of the research-production cycle.[4] Given the commitment and resources devoted to basic and applied research, why is it that the advantages of the Soviet centrally planned system—its capacity to mobilize resources to achieve "mission-oriented" innovations, to train the labor force with the appropriate research and design skills, to acquire foreign technology, to potentially avoid duplication with its centralized determination of innovation; its mandatory enforcement of the introduction of innovation; and its centralized investment and centralized materials allocation—have failed to reduce the technological gap?

Lags are apparent at each stage of the research-production cycle. Soviet economists estimate that *development* activities, which take three to four years in the United States, may take up to six to eight years in the USSR, often causing new equipment to be obsolete by the time it is introduced into production. Despite efforts by Soviet leaders to accelerate R&D—adopting numerous resolutions at Party congresses and plenums and conducting various organizational experiments at the branch and regional levels—official reports appear to concede a reduction in the tangible results of R&D activities.[5] More revealing than the figures documenting a decline in development results, however, is the Soviet report on the economic effect of the fifteen thousand models of new equipment that were introduced in the early eighties: the calculated economic effect—some 150 million rubles per year in cost saving, higher productivity, and so forth—was generated by less than 3 percent of the inventions; *86.2 percent of them produced no return at all.*[6]

Soviet leaders also recognize difficulties with the diffusion of new technologies. According to data provided by the U.S. National Scientific Fund, commercial utilization of an invention occurs, on the average, in 3.6 years in Japan. To bring an invention to commercial utilization in the USSR takes an average of 6.4 years. In their analysis of the lead time between the application for a patent and the introduction of the invention, Martens and Young (1979) find that, at the end of two years, only 23 percent of the Soviet inventions had been implemented, about two-thirds less than the United States and the former West Germany. The magnitude of the diffusion problem is best illustrated in an article in the Soviet press: of those inventions actually implemented (23 percent if Martens and Young are correct), *80 percent were introduced in only one enterprise, less than 20 percent were introduced in three to four enterprises, and only 0.6 percent were introduced in five or more enterprises.*[7]

A variety of barriers to the development and diffusion of new technology have been identified by Soviet and Western scholars alike. These include (1) lack of integration between R&D and production, (2) little incentive for R&D activities, (3) lack of emphasis on development work, particularly preparation of prototypes, (4) lack of competition in R&D work, (5) overspecialization of R&D personnel and agencies, (6) bureaucratic obstacles to collaboration among agencies in different ministries, (7) shortages of experimental and testing facilities —in the machine-tool and instrument building industry, less than 40 percent of the overall number of enterprises have experimental testing/ production facilities—(8) failure by design and development organizations to take into account enterprise-specific technological capacities, (9) duplication of effort and significant refashioning of new technologies to meet enterprise needs. Berliner (1976) offers a systematic analysis of the organizational, pricing, and incentive problems that, combined with the priority to the military, contribute to the technological lag in civilian-sector industries. Of interest in this analysis are the respective roles of planners and managers in the innovation decision and their relative contributions to the growing technological lag.

From the literature, we know that planners' demand for innovation affects both R&D organizations and industrial enterprises.[8] Plan targets reflect planners' demand for innovation, and planners have attempted to devise incentive structures to motivate managers to implement plan targets. Planners have also adopted a price-formation policy whereby innovation increases profits or other plan criteria upon which managerial bonuses are based. Finally, planners have established specific centers for innovation, which, only in some instances, are integrated with production facilities.

We also know from the literature that Soviet enterprise managers have for decades operated in an environment characterized by taut plans, uncertain input supplies, and a discontinuous bonus structure that rewards current production.[9] Facing multiple targets, not all of which can be fulfilled, managers take responsibility for selecting the optimal production strategy. Their decisions, while rational, often generate results contrary to planners' goals. The literature suggests a substantial discrepancy between planners' goals and managerial behavior in the development and diffusion of new technology. This paper proposes an alternative hypothesis—that is, neither planners nor managers unambiguously promoted the development and diffusion of new technology in Soviet industry.

The literature fails to analyze systematically the ways in which both planners and managers have inhibited the development and diffusion of innovation in the USSR. Furthermore, from published sources alone it is not possible to establish a uniform ranking of the dozen or more barriers to technological advance cited in the literature. Nor is it possible to identify ways in which these barriers may interact to further compound the technological lag situation. It is precisely in these areas that firsthand experience by expert informants on the research-production cycle can help us to develop hypotheses, not only about the relative weights to attach to different variables but also about the ways in which these variables interact to impede the development and diffusion of new technology.

Some of the data used in this analysis were collected from in-depth interviews with recent Soviet emigrants to the United States. Participants in the Enterprise Management[10] and Science and Technology[11] interview projects formerly held responsible positions in industrial enterprises, design institutes, scientific research institutes, and other planning, financial, and supply organizations involved in the research-production cycle. Interview evidence contributes to our understanding of the ways in which both planners and managers impede the development and diffusion of innovation in Soviet industry. Interview evidence also offers an "insiders' view" of how innovation decisions are made that is not found in official Soviet sources. Thus, interview evidence is crucial in identifying those issues that must be addressed if perestroika is successfully to reduce the technological lag.

Planners as a Barrier to Innovation

Economic growth has always been a high priority for Soviet planners. In the initial decades after the revolution, economic growth stemmed primarily from the application of more inputs to the production process. The literature suggests that during the extensive-growth environment of the fifties and sixties, planners' demand for innovation focused on technologies that increased the quantity of output, rather than on cost- or resource-saving technologies.[12] That is, the literature portrays planners demanding the construction of new production facilities or expanding existing production areas by increasing the number of machines, rather than renovating or retooling. Indeed, in the opinion of the design engineers who participated in this project and who were reporting on work experiences during the sixties, seventies, and early eighties, creating the ability to produce more was revealed by planners as a high priority by the nature and the number of projects funded. Their descriptions indicate that project-making organizations were not exempt from planners' emphasis on quantity. Like production organizations, research and design organizations faced the "ratchet" principle of plan formulation and plans with "impossibly high targets."

Design engineers said one of the biggest problems at their place of work was the rush to produce designs, with the inevitable consequence of low-quality work. Moreover, since the majority worked in organizations that were not self-financing, they were pressured to respond to planners' demands for specific innovations.[13] Informants state that "the Ministry holds all the cards, it manipulates as it wants"; "new projects had to be completed immediately, projects not initially specified in the plan. This was especially true for defense orders." As one informant reports: "[It was] also hurtful to work on drafts which then just sit on shelves for years and nobody uses them."

The literature suggests that as the availability of inputs fell in the seventies and early eighties, planners' demand for innovation began to focus on cost- and resource-saving technologies, especially on those that simultaneously improved output quality. Yet little evidence is available to support this proposition. One finds, instead, situations where resource- or cost-saving technologies were domestically available, but not widely used.[14] Moreover, despite plan targets for the introduction of new technology, interviews with former managerial personnel suggest that *planners were willing to accept alternatives to innovation*—reducing production reserves, work-time losses, and equipment downtime; increasing the number of work shifts; improving resource utilization; preserving old machinery by extensive repairs—if these alternatives generated the higher output levels specified in the plan.[15]

Interview evidence offers a striking picture of how the various incentive structures designed by planners between 1965 and 1985 adversely affected both the supply of and demand for innovation in Soviet industry. Regardless of whether the informant worked in a design bureau or production facility, *quantity* dominated discussions of the most important plan target to fulfill.[16] Managers and design engineers responded by minimizing the research component within any development program, using already proven parts and subassemblies rather than designing improved components. Managers also report giving conservative estimates of the lead time required to bring a project on line, typically meeting only minimum performance and cost requirements. Because designs tended to focus on simplicity and performance reliability, innovations were at best marginal contributions, not major changes in technology.

That quality was rarely mentioned by participants in the interview project as a primary plan target—one that determined whether or not a bonus would be paid—is hardly surprising.[17] The literature suggests that quality became an important plan target to fulfill only after the 1979 decree, and many of these informants' "last normal period" was coming to an end at that time. Quality did appear as a secondary plan target in discussions about the size of the bonus. In one instance in particular, an informant reports that, when quality targets were met, the bonus was doubled.

Planners acted as a barrier to development of new technology by devoting significant resources to the imitation of existing technology— "reverse engineering." In a striking example, one informant reports difficulties associated with Soviet-produced ball bearings. He had an opportunity to reproduce foreign equipment, which would be used to make more precise ball bearings, but was unable to reproduce this equipment because of the lack of precision ball bearings needed to make the equipment itself. Another informant reports a "request" by ministry officials, first, to install a recently acquired (but apparently not ordered by the informant's organization) foreign machine used in the production of plastic cups and to "have a cup on the minister's desk in three days" and, second, to determine whether similar equipment could be produced domestically.[18]

Reports by informants with firsthand experience with reverse engineering help us to evaluate the type of products or processes that the Soviets may be more likely to reproduce. Interview evidence suggests, for example, that reverse engineering was more likely to occur with those technologies that were not easily protected and in technologies that had the potential to eliminate bottlenecks in priority industries. In some instances where reverse engineering was successful—that is, the technology was successfully imitated—the results fell short of expectations because of the lack of specialized or cooperating inputs. Informants indicate that the availability of hard-currency reserves was more of a constraint influencing planners' decision to purchase the rights to Western designs than the availability of technologically advanced inputs and equipment was in determining whether a given technology would be imitated.

Managers as a Barrier to Innovation

Equally critical as the role of planners in the innovation process is the role of enterprise-management decision making. Soviet enterprise managers regularly decide what materials to use, what level of inventories to maintain, what equipment to order, what schedule to follow for production operations and equipment repair, what goods are to be shipped and when, and whether or not to sue a supplier for breach of contract (Berliner 1976). Thus, to speak only of planners' demand for innovation ignores an important dimension of the introduction of new technology.

Informants describe planning, ministry, and Party officials pressuring them to introduce new technologies, yet at the same time failing to allocate sufficient materials or time to support their efforts.[19] Informants also make it clear that introducing innovation was not a primary plan target. At best, introducing new technology affected the size of their bonus, but that component of the bonus was always reported to be rather small. Moreover, informants repeatedly underscored the fact that, for them, penalties for failing to innovate were never very great. Failing to fulfill the plan for the introduction of new technology represented at most only a 2 or 3 percent reduction in the bonus. Thus, from the manager's perspective, there was no real reward for fulfilling this aspect of the plan and very little penalty for failing to do so. Under such conditions, managers were more inclined to pursue alternatives to innovation that generated the same output level or bonus outcome than to adhere to the plan to introduce new technology.

Interview evidence suggests that management's willingness either to respond favorably to the pressure from above or to initiate innovation is determined by (1) *personal characteristics* (that is, whether the manager is a "good soldier," "risk taker," "progressive leader"), (2) *objective conditions* (availability of necessary financial and physical resources, adequately skilled work force, impact of innovation on future plan targets), and (3) *subjective conditions* (sufficiently high bonus payments, advancement along a career path). It is here that the link between technology, culture, and development is most evident.

Cultural factors are more likely to influence personal characteristics and subjective conditions than they are to influence the objective conditions surrounding the development or diffusion of new technology. Indeed, informants' discussions of the objective factors influencing innovative activities were dominated by references to (1) the importance of access to new technology (especially foreign technology), (2) the cost (including acquisition, installation, and operation cost) of the new technology, and (3) the quality of the new technology in terms of productivity, durability, reliability, and savings. Informants were unanimous in the opinion that access to new technology depended upon the organization's proximity to the defense industry and to whether the organization earned foreign currency.[20]

In a recent article pertaining to the influence of social and cultural factors on personal characteristics, Baumol (1990) argues that the supply of entrepreneurs in a society is fixed at any point in time and that policy—by shaping economic, political, social, and legal institutions and by establishing the relationship between risk and reward—determines whether entrepreneurs will engage in legal or illegal activities. While perestroika may not as yet have permeated the Soviet culture in such a way as to generate an increase in the number of entrepreneurs, it has caused those with entrepreneurial talents to move into cooperatives and joint ventures. Despite their somewhat tenuous legal status, and despite bureaucratic resistance to and the lack of popular support for cooperatives and joint ventures, their numbers have grown rapidly (Tedstrom 1990). Policies that legalized and thus reward activities for collective, if not individual, gain are influenced by cultural factors. To the extent that policy determines the relationship between risk and reward, policy impacts on the subjective conditions influencing innovative activities. Indeed, Gustafson (1990) argues that cultural factors are undermining the possibility of Soviet leaders adopting policies that will legalize and reward activities for personal gain.

Soviet managers have clearly been capable of evaluating both the risks and rewards associated with introducing new technology in their enterprise. In an environment of taut plans, uncertain input supply, and a bonus structure that provided relatively high rewards for fulfilling current production targets and minimal rewards for innovation, managers tended to act as barriers to innovation. This is especially true in light of planners' reluctance to reduce plan targets to compensate for production delays arising from installing new technology.[21] Interview evidence suggests that management's reluctance to innovate would have been greatly diminished if plan targets had been reduced to allow for retooling or changeover.

Prominently featured in managers' discussions of innovation are difficulties associated with obtaining the necessary inputs required by the new technologies. In many instances, innovation required establishing relationships with new suppliers, which, given pervasive transportation bottlenecks, meant that geographical location was a crucial factor in timely deliveries. Managers felt that planners ignored location when assigning suppliers and, thus, fought to avoid changing suppliers with whom good relations had been established. Moreover, enterprise managers facing a sellers' market—that is, persistent shortages of a broad array of necessary inputs—frequently elected to self-supply as many components as possible. Thus, adopting a new technology not only jeopardized the availability of inputs from assigned suppliers but also jeopardized the firm's ability to self-supply.

Informants reported that at their place of work the price, or cost, of new process technology was calculated as a function of its "use value"—where the usefulness of the process technology was defined by its impact on productivity, current production costs, and so forth. In their discussions, it was evident that the supplier of the process innovation had the responsibility for providing the use-value calculations; that is, the design bureaus and scientific research institutes calculated the economic effect to be realized by production facilities with the introduction of the new technology.

Interview evidence suggests a number of interesting propositions regarding these calculations. First, when bonuses depend upon the magnitude of the calculated economic effect, innovating organizations have little incentive to provide reliable computations. Given a reward structure in which higher estimates result in larger bonuses, design institutes benefit from overestimating the economic effect. Informants formerly working in design bureaus and those employed in client organizations describe problems related to computational reliability. Second, because of the way prices are determined for output produced by enterprises introducing the process innovation, enterprises are forced to cover any additional costs that arise when the estimated and actual "savings" diverge. Interview evidence provides a wealth of data related to estimated and actual project costs, why they differ, and how suppliers and clients respond to these situations. Third, the pricing structure is such that enterprises must also cover the cost of installing new technologies—technologies that may or may not meet enterprise-specific requirements. Informants describe in detail the time and documentation necessary to obtain special funds from the ministry to refashion the technology to suit enterprise needs. Managers frequently related instances where even sufficient financial resources failed to solve installation nightmares. In the sellers' market environment that production managers face, there is little recourse when technologies fail to perform as specified. The client must bear the entire cost of "defective" innovations.

Installation difficulties, supply difficulties, and performance discrepancies all contribute to the risk of introducing new technology. Risk is only a problem, however, when it is not offset by adequate rewards. An incentive structure that fails to compensate for the risk of innovation represents a major barrier to the introduction of new technology.

Soviet Innovation Environment:
A Summary Statement

Interview evidence suggests that both planners and managers impeded the development and diffusion of innovation in Soviet industry. Planners instituted a bonus structure based on the fulfillment of current production targets with little reward for the introduction of new technology. They routinely set plan targets high relative to the firm's productive capacity, annually raising plan targets without corresponding increases in inputs and without allowing for the downtime required to retool for a new product or process technology. Planners had monopoly control over the resources allocated to innovation and elected to devote a significant share to the acquisition of foreign technology, regardless of the domestic availability of cooperating inputs. Furthermore, planners' control over innovation decision making imposed substantial documentation requirements upon managers, prolonging both the development and diffusion of new technology.

Prior to the 1990 economic-reform proposals, managers were more inclined to pursue alternatives to innovation that generated the same output level (bonus) as that specified in the plan for introducing new technology. Regardless of whether managers were responding to pressure from above or initiating innovation, they routinely considered access, quality, savings, and cost. Informants identified factors influencing access to innovation and indicated that management's reluctance to innovate would be greatly diminished if plan targets were reduced or management was otherwise compensated for the inevitable production delays during the installation of new technology. Managers uniformly acknowledged that bonuses for introducing new technology were ineffective because they failed to offset the risks involved. A major risk was associated with the sellers' market environment where managers were unable to enforce their requirements on design specifications and were obliged to cover any costs associated with refashioning the technology.

While space constraints prohibit a detailed reporting of interview evidence, the inescapable conclusion from the informants' descriptions of the Soviet innovation environment confirms Thornton's findings that (1) because R&D is usually separate from production, innovation in the Soviet economy sometimes is inappropriate for the user, with long lags from research to production; (2) because of the lack of property rights over inventions, Soviet innovators have no incentive to extend the sphere of application; (3) because new technology is often acquired piecemeal, and without cooperating inputs, foreign equipment is less productive in Soviet use, and new technology often falls short of predicted results; (4) slow diffusion of innovation in Soviet industry stems both from perverse incentives for the adoption of new technologies and the lack of structural flexibility in the economy; and (5) innovation is impeded by the poor service and support for new equipment after delivery (Thornton 1988).

What are the prospects for improving the innovation environment in the USSR? The prospects are significantly brighter if perestroika succeeds in dismantling the centralized resource-allocation mechanism in the Soviet economy. However, Soviet planners and managers have been trudging on a "treadmill of reform" (Schroeder 1979) since the mid-sixties, seeking ways to reorganize institutions and revise success criteria, while still maintaining a centralized decision-making structure. Each step on the treadmill was lauded as a necessary measure to overcome technological lags and acquire the ability to develop and diffuse innovation more readily. Gorbachev stepped onto the reform treadmill in 1985, initially pursuing conventional tactics to improve economic performance—calling for more effort and less waste. More radical reform elements were introduced under the guise of glasnost and perestroika only after it became evident that economic performance had failed to improve or even to stabilize.

Five years have passed, and the Soviet leadership is only now beginning to address seriously the fundamental issues required to bring glasnost and perestroika to successful fruition. Unfortunately, five years of sporadic, contradictory, incomplete, and perhaps inept reform proposals have created a crisis in the Soviet economic and political systems. In such an environment, public discussions of innovation and future production will no doubt be postponed until more immediate concerns are resolved. It is impossible to imagine, however, the adoption of reform legislation that would seriously undermine the creation of an environment more conducive to innovation. Thus, while discussions of future production are politically unwise in a situation of current shortages, Soviet leaders are unlikely to adopt either the institutions or the policies that would further impede the development and diffusion of new technology.

Transition to a Market Economy

Confronted with the unraveling of the economic and political bureaucracy, the Soviet people appear more willing to consider seriously the adoption of emergency measures to initiate a break from the centrally administered economic system and begin the transition to a market-oriented economy. It is clear that Shatalin's 500-day program is not intended to create a market economy in the USSR. Rather, if implemented—even if it takes more than the proposed five hundred days—it will significantly diminish the dominant role of centralized resource allocation and thus permit more decentralized decision making to take place.

At issue here is the extent to which the program will change the economic environment in such a way as to impact on planners' and managers' innovation decisions. If the transition succeeds and a market economy is allowed to develop, profitability considerations will require Soviet managers to reevaluate their strategies regarding current and future production. And planners will increasingly need to rely on economic levers—taxes and subsidies, for example—to influence current and future production.

Two transition proposals are currently on the table: Shatalin's and Gorbachev's. They differ in nature and pace and, thus, generate different consequences for innovation decision making.

Shatalin's 500-Day Plan

The thrust of the proposal, "Transition to the Market: Concept and Program," drafted by a working group formed in August 1990 by a joint decision of Mikhail S. Gorbachev and Boris N. Yeltsin and now popularly called the "Shatalin 500-day plan," is to create the conditions necessary to initiate the transition to a market economy. The focus of the plan is on abolishing the institutional framework that sustains the command economic structure, establishing a monetary system that will facilitate currency-convertibility and economic-stabilization policies, eliminating state control over the majority of prices, and dividing decision making between republic and central authorities. The ultimate goal of the 500-day plan is to privatize some 46,000 industrial enterprises and 760,000 trade organizations by the end of the transition-initiating period.[22]

The first phase of the transition-initiating period hinges on the successful creation of institutions to implement the necessary monetary and fiscal policies to limit inflationary pressures in the economy. The second phase calls for a widespread price reform and the demonopolization of the economy.[23] It is significant that Shatalin's proposal calls for each republic to decide on the pace of reform. That is, in the division between republic and central authorities, republics will have the right to decide on the nature and scope of property rights,[24] the regulation of incomes and social security, and on appropriate investment activities. The center will have control over monetary and fiscal policy (with special rights to adopt "extraordinary measures" for limiting the growth in the money supply and overcoming budget deficit), foreign-currency policy, price-formation policy, and the reorganization of foreign trade.

Shatalin's program identifies four periods in the process of creating the conditions necessary for the transition to a market economy. In each period, specific actions are outlined. For example, in the first one hundred days, the transition-initiating program calls, first, for the legalization of private property and individual activity for personal gain; second, for the selling or granting of state-owned dwellings, land, and physical assets to individuals, collectives, or joint-stock companies; and, third, for the granting of emergency powers to the president to reform the monetary and fiscal systems in order to stop the growth in the money supply and to reduce state expenditures. More specifically, centralized expenditures are to be reduced on (1) foreign aid, (2) defense (10 percent), (3) KGB (20 percent), (4) enterprise subsidies (30 to 50 percent), and (5) investment (20 to 30 percent). Moreover, investment projects with budgets exceeding 100 million rubles are to be automatically rejected. To finance existing expenditures, taxes (turnover tax, profit tax, personal income tax, and a newly created tax on nonrenewable resources) are to be raised, and bonds issued. Funds saved from reducing enterprise subsidies and closing loss-making firms are to be put into a stabilization fund to aid firms and workers during the transition period. To offset supply disruptions from plant closings, firms previously shut down because of environmental concerns are to be reopened. During the first one hundred days, a wholesale trade network is to be established, and prices, except for those on goods produced under state order, are to adjust to surplus and shortage conditions. The plan makes it clear that state control over prices for goods "of prime necessity" are to remain in place. Citizens are to be granted the right to own Western currency. To handle the existing cash overhang in the Soviet economy (Linz 1990), incomes are to be indexed, interest rates on savings deposits are to increase, and the supply of consumer goods and services is to continue to grow as a result of the conversion of defense industries to civilian production, the sales of state property, and the availability of newly constructed residential housing, garages, and dachas.

Controlling inflation while removing state control over production and distribution is the main focus of the second stage of the program (days 100–250). Shatalin's program gives republic and local authorities the authority to determine the pace of market-based price formation and calls for a public agreement on wage control. While state orders (*goszakazy*) are to be completed, some 1,000 to 1,500 joint-stock companies are to be created from the sale or dissolution of large state enterprises. During this stage, up to 50 percent of the small retail and catering organizations are to be privatized.

In the third stage, days 250–400, the focus is on stabilizing the consumer market. This is to be achieved in part by the sales of housing, in part by wage control, and in part by setting aside some 3 to 7 percent of the production of consumer goods to distribute to low-income families or to be otherwise rationed by central or local authorities. By the end of the first four hundred days, state control over prices is to be removed from 70 to 80 percent of the domestically produced goods; state-controlled prices are to be maintained only for primary resources (crude oil, oil products, gas, and some kinds of ferrous and nonferrous metals) and some consumer goods (bread, meat and dairy products, medicine, textbooks, public transport, and utilities). Some 30 to 40 percent of the industrial capital assets, up to 50 percent of the construction organizations and motor-transport facilities, and not less than 60 percent of trade, catering, and service organizations are to be sold to private, collective, or joint-stock companies.[25]

In the last one hundred days of the transition-initiating period, "antimonopoly industrial restructuring" is to continue. To facilitate and protect domestic industry, Shatalin's program calls for (1) abolishing the passport system in an effort to enhance labor mobility, (2) establishing custom tariffs to reduce foreign competition, and (3) promoting performance-based pay and profitability/rate-of-return calculations for investment decision making.

Several aspects of Shatalin's program will affect the innovation environment. Most important are those elements that will allow for entry of firms: legalizing private property and individual activity for personal gain; eliminating subsidies to loss-making firms and selling the physical assets of those obliged to shut down; and establishing capital markets where interest rates influence the supply of and demand for loanable funds. Perhaps equally important are those elements that will result in prices responding to surplus and shortage conditions. Unfortunately, the proposal retains state-controlled prices for energy and other basic raw materials, transportation, communication, and many basic consumer goods. While wage controls may help to dampen inflationary pressures, they may also effectively restrict labor mobility between occupations and geographic regions.

Somewhat surprising is the prominence attached to converting defense plants to civilian production as a means of quickly upgrading product quality. It is rather unlikely that defense plants can be converted quickly—it took fifteen years after World War II—and even more unlikely that any productivity differential can be sustained in the absence of priority deliveries of the highest quality inputs (Tedstrom 1990).

Gorbachev's Compromise

The commitment to initiating the transition to a market economy is not as evident in the compromise program outlined by Gorbachev.[26] State control over production, distribution, and prices is to be phased out much more slowly than proposed by Shatalin. Not until 1992, for example, are price controls to be removed, and even then they are to be maintained on "essential consumer goods" (bread, meat, milk, and so forth) and "key types of raw materials, production output, goods, and services." Central control is to remain over the production and distribution of oil, gas, gold, diamonds, and other natural wealth and over key sectors of the economy: transportation, communications, defense, energy, credit and monetary policy, and foreign-trade policy. Subsidies to inefficient firms are to continue indefinitely, with only the "hopelessly inefficient" firms being closed.

The compromise underscores Gorbachev's position that "only under conditions of a strong and clearly organized state power" can the transition to a market economy succeed. Because the compromise reduces

the scope and slows down the pace of the transition process, it pro-
longs the period before decentralized decision making is actually implemented. As such, the compromise proposal may undermine the
likelihood of a successful transition. However, given the strong and
widespread support by proponents of the transition in the various republics, the probability that the command economy will be sustained is
rather low. Thus, the operating assumption of this paper is that the
transition to a market-oriented economy will succeed and that innovation decision making in the Soviet Union will differ markedly from
past experience.[27] It is assumed that managers who survive the transition period are those whose skills are more than adequate to meet the
challenges inherent in a decentralized decision-making environment. It
is also assumed that, at a minimum, planners will be employed to
coordinate monetary and fiscal policies.

Innovation Strategies: Planners and Managers

Innovation decision making by Soviet planners and managers in the
transition and post-transition economy will necessarily be guided by
the same factors that guide innovation decision making in market-oriented economies.[28] Managers in the short term will be motivated to use
existing assets as efficiently as possible, in the medium term to substitute between capital and labor to produce at least cost, and in the long
term to develop new products and processes that open new markets or
restructure old ones. Soviet managers, like their Western counterparts,
will be obliged to develop a global perspective and thus adopt financial
and accounting practices, product design and quality standards, and
market research and marketing strategies that will enable them to compete effectively in world markets. Planners, like policymakers in market economies, have the option of employing taxes and subsidies to
influence innovation decision making and thus stimulate economic
growth and consumer welfare.

   To facilitate the comparison between planners' and managers' strategies regarding technological advance, the analysis here will be restricted to product innovation, where products are differentiated on the
basis of quality, and quality differences are represented using a ladder
analogy—each rung on the ladder is associated with a higher-quality
product.[29] Resource constraints force managers to choose between developing new products, adding a new rung to the ladder, or copying

existing products, moving up the existing rungs. Developing new products is more risky, but more likely to bring about greater profits than copying existing products.[30] Firms will innovate when the benefits of adding a new rung to the quality ladder exceed the benefits associated with profit sharing. Imitation is more profitable when the technological gap between firms is large, that is, the greater the distance between rungs on the ladder. Moreover, the larger the number of rungs from the top to the bottom of the quality ladder, the more likely are firms on the lower rungs to imitate.[31]

Both the development of new technology and the diffusion of existing technology contribute to economic growth.[32] Planners adopt policies to promote economic growth and enhance consumer welfare. Given resource constraints, when firms decide to invest in the development of new products for future production, current production falls below what it otherwise could have been, and consumer welfare, to the extent that it is based on current consumption, also falls. To the extent that consumers prefer new products to old, welfare will rise with the introduction of the higher-quality product that results from the investment in innovation. When firms decide to imitate existing products, resources cannot simultaneously be used to develop new products. However, imitation brings firms closer to state-of-the-art production techniques and consequently contributes to economic growth. Because imitation is less costly than innovation, current production and thus welfare fall by less, but no new, higher-quality product is available to consumers to enhance their welfare.

Planners, facing these consequences for growth and welfare, must differentiate their policy in accordance with the firm's strategy regarding innovation or imitation. That is, focusing only on the development, or only on the diffusion, of new products will generate lower growth rates and lower consumer welfare than would arise as a result of adopting policies that respond to industry-specific conditions. Segerstrom (1990) demonstrates that the optimal strategy for planners is to subsidize innovation when all firms in the industry share the same technology and to subsidize imitation when there is a significant technological gap between firms in an industry.

Will Soviet managers innovate or imitate? Numerous studies on the technological level of Soviet industry indicate that product quality for many goods is relatively low. Because Soviet managers are starting out rather far down on the quality ladder, imitation will no doubt be the dominant strategy during the transition and the initial post-transition period. As Soviet firms move up the quality ladder, managers confront the choice of innovation or imitation as the appropriate strategy.

Under what conditions will Soviet managers elect to invest in innovation? Firms are motivated to innovate when benefits exceed cost. Benefits are measured by the profits realized from the new product. Not until another firm successfully imitates is profit shared. Consequently, firms are likely to innovate if the time lag prior to imitation is sufficient to generate a profit stream that more than covers the cost of the innovation.[33]

Under what conditions will Soviet managers elect to invest in imitation? Products that can be analyzed fairly easily, products that are sold to buyers who are more concerned about quality and technical specifications than price, and products that do not require costly retooling of production processes are the most likely candidates for imitation (Baldwin and Childs 1969).

The literature establishes a more general set of conditions regarding the innovation–imitation choice (Cheng 1989, Baldwin and Childs 1969, Jovanovic and MacDonald 1988, Reingaum 1982, Segerstrom 1990). As Segerstrom (1990) explains, at any given point in time, firms either are located in an industry dominated by a single quality leader or in an industry with two quality leaders. Managers of firms located below the top rung of the quality ladder must choose between developing a new product or copying the existing product produced by the firm or firms on the top rung of the quality ladder. Firms in an industry with two quality leaders will choose to develop a new product because copying simply adds another firm to the profit split and results in a situation where benefits are low relative to cost. Thus, in industries where product technology is already diffused across firms, management's optimal strategy is to invest in innovation.

Successful innovation adds a new rung to the quality ladder, with the innovating firm enjoying monopoly profits. In an industry with only one quality leader, firms are more likely to imitate than to innovate. The quality leader has less to gain from innovating than from waiting until imitation occurs and profits disappear before spending money on R&D activities.[34] Segerstrom (1990) shows that firms one rung below the quality leader are more likely to imitate rather than innovate because of the relatively lower costs and sizable profit share. Successful imitation transforms the industry from one with a single quality leader to one with two quality leaders, creating an environment where innovation once again is management's optimal strategy.

Planners can influence managers' choice between innovation and imitation by electing to subsidize these activities. To maximize the impact of the subsidies on economic growth and consumer welfare, differential policies are required. That is, if the industry is characterized by two quality leaders, Segerstrom (1990) demonstrates that the optimal policy is to subsidize innovation. Innovation subsidies lower the cost of innovation. Facing a lower cost, more firms are likely to invest in innovation. The greater the number of firms investing in innovation, the higher the probability that innovation will occur. Once a new product is developed, the rate of imitation will rise as firms devote resources to copying the higher-quality product in order to share in the profits. Given fixed resources, an increase in the rate of imitation must result in fewer resources devoted to innovation. However, the diffusion of product technology contributes to economic growth, more than offsetting the reduced rate of innovation.[35]

While this diffusion of technology augments growth, the impact on consumer welfare is ambiguous. Innovation subsidies lower the cost of innovation and result in a bias toward overinvestment in innovative R&D. If more resources are devoted to R&D rather than current production of consumer goods, consumer welfare falls. The decline in welfare may be offset in the next period if economic growth generated by R&D activities is sufficiently high.

A similar caveat applies to imitation subsidies. Resources devoted to imitation cannot simultaneously be used for innovation or current consumption. On the one hand, investment in imitation reduces current consumption and resources available for the development of new products in the future (but by less than that required by innovation) and, consequently, reduces consumer welfare. On the other hand, imitation hastens and broadens the availability of the highest quality goods and thus increases consumer welfare. Subsidizing imitation would be the appropriate government policy for firms in industries with a single quality leader. As subsidies reduce the cost of imitation, more firms would be quicker to imitate, shortening the time period that any one firm sustains its profit share. As firms imitate, profits would necessarily be split among a larger number of firms. Knowing this, profit-maximizing firms in a single-quality-leader industry will respond to the imitation subsidy by devoting resources to innovation. As the in-

tensity of innovative effort increases, in an environment of fixed resources, resources devoted to imitation must fall. Because it takes
more resources to innovate than to imitate and because the diffusion of
product technology necessarily falls, the impact of imitation subsidies
on economic growth and on consumer welfare is ambiguous. Consumer welfare rises with the development of a higher-quality product,
but a larger share of current resources is required to obtain it.

Conclusions

Graham (1990) has carefully documented the fits and starts of Russian
and Soviet technological advance, offering several explanations for the
uneven pattern of the development and diffusion of innovation. Josephson (1990) analyzes the implications of perestroika for technological development. This paper identifies those elements of the Soviet
innovation environment that are likely to change as a consequence of
perestroika and how planners and managers are likely to revise their
decision making if the transition to a market-oriented economy succeeds. If perestroika succeeds in reducing repressed inflation, introducing flexible pricing, and encouraging competition, Soviet planners and
managers will face a radically different innovation environment.[36] The
empirical and theoretical evidence presented in this paper suggests that
cultural factors are unlikely to represent barriers to innovation and,
thus, will not impede Soviet managers in this new environment.

Notes

1. *Perekhod k rynku. Kontseptsiia i programma. Chast' I* (Moscow:
Arkhangel'skoe, 1990).
2. The Soviet Interview Project was supported by Contract No. 701 from the
National Council for Soviet and East European Research to the University of
Illinois at Urbana-Champaign, James R. Millar, Principal Investigator.
3. Frequent references in the Soviet literature address the continued failure to
fulfill plans to introduce new technology. According to Soviet economists, of the
unfulfilled assignments for new equipment, 20 percent arose because ministry
officials failed to transmit annual plan targets in a timely manner, 15 percent
resulted from delays in deliveries of materials by subcontractors, 28 percent were
the result of construction delays (including failure to allocate sufficient financial
and physical resources), and 33 percent stemmed from organizational problems
involved in implementing new technologies (see *Pravda*, 6 May 1982, p. 1). The
failure to close the technological gap is most strikingly portrayed in Berliner's

(1988) examination of the wording of speeches by Malenkov in 1941 and Gorbachev in 1986 regarding the failure to sustain the development and diffusion of new technology.

4. See, for example, Amann, Cooper, and Davies (1977); Amann and Cooper (1982); Berliner (1976); Boretsky (1966); Cohn (1976, 1979); Gavrilov (1975); Hill and McKay (1988); Kushlin (1976); Martens and Young (1979); Nolting (1976); and Sutton (1973).

5. From 1960 to 1965, for example, some 4,700 models of new equipment were developed in the USSR; from 1966 to 1970, 4,300 new models were developed; from 1971 to 1975, only 4,000; and from 1976 to 1980, just 3,600 (Zaychenko 1988). The reduction in development results is not a consequence of the elimination of R&D duplication across branches.

6. For further discussion see Linz and Thornton (1988).

7. "Getting Research Gains into Production," *Current Digest of the Soviet Press*, vol. 37, no, 5 (1985): pp. 14–15.

8. Even those design bureaus and scientific research institutes that are self-financing (*khozraschetnyi*) have to date received detailed plans from their ministry. The enterprise annual plan includes specific requirements for innovation, including (1) directives for automation of production processes and the introduction of advanced technology; (2) directives for development and production of prototypes; (3) directives for the most important research, development, and experimental projects; (4) a list of obsolete machinery and equipment whose production is to be terminated; (5) directives for quality of production of new kinds of industrial products; and (6) a statement of requirements of materials, supplies, equipment to implement this plan (Grossman 1965).

9. For a survey of the recent literature, see Linz (1988).

10. Description of the Enterprise Management Project is found in Linz (1986).

11. For a complete description of the Science and Technology Project, see Thornton and Linz (1988) and Linz and Thornton (1988).

12. See, for example, Amann, Cooper, and Davies (1977); Bergson (1978); Bliakhman (1968); Cohn (1976); Green and Levine (1977); Hanson (1976); Nolting (1976); and Popov (1976).

13. Informants who formerly worked in R&D organizations report that most of their activities were specified in work orders drawn up by a superior organization —usually a ministry. The larger the size of the project, the more levels of ministerial hierarchy had to review and approve the project. These activities were initiated with the signing of a contract called a *tekhnicheskoe zadanie* (technical assessment), describing a scientific-research project (*nauchno-issledovatel'skaia rabota*), and the subsequent experimental-design project (*opytno-konstruktorskaia rabota*). The terms of these technical assignments specified expected results and work schedule, the planned level and structure of staffing, compensation as well as incremental benefits for early completion or above-plan performance, and penalties for late or deficient performance. See Thornton and Linz (1988).

14. Granick (1988) suggests that labor-saving technology has not been a high priority in the Soviet Union, citing many instances where such technology was not put in place when available. Leary and Thornton (1989) find that although continuous casting was introduced in the Soviet steel industry in the fifties, it has not to date been widely used, despite the fact that it produces less waste. It should

be noted that the U.S. steel industry was slow to introduce two proven cost-saving technologies—oxygen furnace and continuous casting. See Adams and Dirlam (1964).

15. Interview evidence corroborates the recent Soviet literature describing alternatives to innovation as a means to increase output. See, for example, Arakelyan (1987) and Arbuzov (1987).

16. For research and design organizations, completing a project on time and not overspending the budget were also important indicators. To complete projects on time and to avoid overspending the budget, leaders of research and design organizations described not only their efforts to incorporate large, high-budget projects into the plan but also their unofficial shifting of resources and personnel between projects. Interview evidence thus offers insight into planners' and managers' decisions and actions that have to date resulted in a high level of R&D expenditures in the USSR without a correspondingly high rate of return. Neither planners nor managers calculated the return on investment. In organizations that received their entire budget from the ministry or were subsidized by state funds, completion not cost dominated the reward structure. Much more attention to cost and resource utilization was paid by organizations that were *khozraschetnyi*.

17. The absence of quality standards comparable to world standards did not go unnoticed by design engineers in this study. One informant reports: "In our industry [machine building], we definitely were behind in production quality. We were always behind in introducing the latest equipment. We knew what was elsewhere, but it just started to appear in the Soviet Union, much later [than it was adopted elsewhere] and [what we had was] of poorer quality." Another informant states that "We didn't always use what we knew to be the best available technology, the newest equipment, because sometimes there was no money to buy that equipment. So we installed what we had. For example, in the Soviet Union we didn't make very precise ball bearings. If we had the money, we bought Swedish ball bearings. Soviet-produced ones were much worse, but we used them when we had to." Yet despite their recognition of the relative backwardness of Soviet technology, these design engineers reported that at least three-quarters of their work in the seventies and early eighties involved modifications of existing Soviet technology. Only those formerly working in such "high-tech" sectors as computers and biotechnology, and in such organizations as experimental-design and production facilities, reported the reverse situation. For further discussion, see Linz and Thornton (1988).

18. Informants report that "We based one of our projects on a German polishing machine tool. We wanted to copy it exactly, but since we had [domestically produced] different parts, we had to change [the design] a little." Another states "We were familiar with Western types of equipment. We had access to [devices] purchased in the U.S. . . . it was bought as an example of the kind of technological achievements available in the West. There were devices we copied, reproduced exactly like the one purchased, but it worked worse because the technology used to produce it was not the same, it was inferior." A third reports, for example, "I was involved in an all-union construction project in which a new machine was being launched [introduced], and for it equipment was received from West Germany. They designed an enormous knife, guillotine type. We [USSR] received that equipment, but the knives that work on that machine were not purchased—

someone made a mistake. So they [ministry officials] decided to economize [make do with what was available in the USSR]. It was necessary to sharpen the knife, but it was not possible to do so using the methods that existed in the Soviet Union at that time. To repair that machine turned out to be impossible and the ministry gave us the task to manufacture such a machine. We started from its documentation, searched for new details, and that machine was made and worked perfectly."

19. One informant reports that "Factories [managers] must endure a lot. If the enterprise has a history of enduring a lot, the Ministry will continue to overload it even more, until the factory [manager] just gives up. In our factory, that situation happened less frequently than with other factories. In a neighboring factory, the plan targets were raised so high that it was impossible for the plant to fulfill the plan. The high targets were not backed up with sufficient materials and to do more with the same equipment, same people was not possible. . . . The director was replaced, plan targets were significantly reduced, and the factory started fulfilling the plan. . . . They overload managers like horses and work them until they drop. . . . Ministry gets higher targets each year from government and must distribute those higher targets among factories. Somebody has to fulfill those targets. Ministry says 'You have to do it. If you don't you will be punished.' " Another informant states "They [party officials] participate only in words, trying to pump us up. . . . If plan targets are not being fulfilled, director is called to gorkom and they 'inspire' him . . . but they don't have any means to help—only vocal inspiration, only orders."

20. Design engineers report that they avoided using foreign equipment or parts in any projects they designed because it was a sure way to guarantee that the project stayed on the shelf. One informant reports: "Whether we included foreign equipment in our designs depended on how much money our client had for the project." A second states: "When we supplied machines to Japan, they stipulated that all electrical equipment used in our designs had to be American or German. Otherwise, they wouldn't take it." A third says: "In my work, I tried to avoid using foreign equipment as much as I could. Using it would inevitably lead to the project getting killed because it was impossible to acquire foreign equipment."

21. Our interview data contain many examples in which modernizing enterprises were expected to simultaneously retool an existing production line and continue previous levels of production—normally inconsistent activities. A number of cases were also detailed in which new products or processes were deficient in important ways. An extreme example is the case reported by a designer in the motor-vehicle industry in which a new model could not be assembled once production started because various components could not be mass-produced with the required tolerances that had been achieved by handcrafting when the prototype was built. In another case, a newly installed unified assembly line was indeed more productive when running than the several separate machines that it replaced. But, each time a part needed replacement or a component needed adjustment, the whole assembly line had to be shut down. Before modernization, production continued on the remaining machines when any one machine was stopped for repair or readjustment (Thornton and Linz 1988).

22. According to data presented in the program document, if the state sells organizations employing 0 to 7 individuals, more than 500,000 employees will leave the state sector. This would result, for example, in 68 percent of the food

stores no longer under state control. If the state sells organizations employing 7 to 10 individuals, more than 77,000 retail stores and some 700,000 employees will no longer be working in the state sector. State control is to be maintained only over that part of retail trade that handles rationed goods. The program also calls for the state to sell small cafes, those with a seating capacity of fifty persons. This would involve 50,000 catering enterprises, thus about 60 percent of all customers served and about 60 percent of total catering sales would be handled by the nonstate sector. In addition to the retail trade and catering organizations, the goal of the 500-day transition-initiating plan is to permit some 20 percent of the industrial enterprises to leave the state sector by the end of 1990, 60 to 70 percent to be sold off by the middle of 1991, and a total of 80 percent to be "denationalized" by the end of 1991. Prior to the sales of industrial and service sector organizations, the state would need to inventory assets and evaluate their value. The speed with which this is to be done in Shatalin's plan has all the features of a fire sale. The inevitable low prices (Brada 1988) may be instrumental in attracting foreign capital.

23. Shatalin's plan calls for the legalization and support of free enterprise: declare amnesty for those sentenced for entrepreneurial activities, adopt laws to protect property and investments, establish uniform accounting and bookkeeping procedures. It should be pointed out, however, that even this "radical" plan restricts the gains individuals can realize from the use of private property. For example, apartments and land would be sold but only on the condition that they would not be resold—except to the state.

24. Central authorities will retain responsibility for denationalization of aerospace, instruments, communication facilities, electronics, shipbuilding, ocean fleet, seaports. Republican and local authorities will be responsible for denationalizing trade, services, catering, agriculture, and small industrial enterprises. In both instances, the procedure calls for central or local authorities to publish a list of eligible firms and to wait one month for the submission of purchase proposals. Some sectors will be exempt from denationalization: defense enterprises, nuclear power and other enterprises employing nuclear technologies, pipelines, long-distance communication networks, KGB and Defense Ministry installations, major railways, and some other railways. During the transition period, the state also is to control the postal and telegraph services and the production and distribution of energy.

25. It is expected that the Soviet people will ultimately spend some 50–60 billion rubles in purchasing the financial and physical assets sold by the state during this 500-day transition-initiating period.

26. Excerpts from "The Main Guidelines for Stabilization of the Economy and Transition to the Market" were published in the *New York Times*, 17 October 1990.

27. A recent example of how the introduction of market institutions continues despite the absence of a consensus over the pace and extent of reform is reported by Elisabeth Rubinfien in the *Wall Street Journal*, "Americans Help Design Soviet Grain Exchanges" (10 December 1990).

28. As long as price controls and centralized allocation remain in place for a significant portion of primary inputs and strategic manufactured goods, decisions made using the available biased information will result in suboptimal outcomes. If these firms attempt to compete in world markets, this will contribute to a higher-than-average failure rate.

29. For a recent survey of the quality-ladder literature see Cheng (1989).

30. Empirical studies of innovation (Scherer 1970, Rogers and Shoemaker 1971, Adams 1961, Davies 1979, Griliches 1979, Sahal 1981, Gort and Klepper 1982, Dorfman 1988) focus on the factors contributing to the cost of innovation, the relationship between innovation and market structure and innovation and firm size. One finds in the theoretical literature three basic models: (1) models where individual firms act as perfect competitors who undertake R&D to reduce costs (Smith 1937, Nelson 1982, Jensen 1982, Jovanovic and Rob 1987), (2) models where firms obtain strategic advantage by improving production processes (Scherer 1967, Kamien and Schwartz 1982, Telser 1982, Spence 1984), and finally, (3) models analyzing where new ideas come from and how they spread (Schumpeter 1934, Salter 1966, David 1969, Futia 1980).

31. Empirical studies (Mansfield et al., 1982) show that imitation costs less than innovation—only about 65 percent of the cost of innovation—because firms are less likely to make mistakes; the follower learns from the leader. Moreover, because marketing and promotion costs have already been expended by the innovator, the imitator acts as a free rider. Finally, firms that have cost advantages resulting from better production facilities, economies of scale, skilled work force, access to better materials, or better quality control have a strong incentive to let another firm undertake the R&D expenditures required to develop a new product and then exploit their own lower production costs in order to realize higher-than-average imitation gains. Empirical studies also show that imitation takes less time than innovation and that patents are not a barrier to imitation, they just raise the cost (Mansfield et al., 1982). The theoretical literature focuses on the choice between innovation and imitation (Cheng 1989, Janovic and MacDonald 1988, Jaffee 1986, Dasgupta 1988, Segerstrom 1990, Reingaum 1982, Baldwin and Childs 1969, Grossman and Helpman 1989, Jensen and Thursby 1986, 1987).

32. The importance of technological advance in both the pattern and determinants of economic growth is spelled out in Solow (1957, 1988), Griliches (1963), and Jorgenson and Griliches (1969). For the purposes of this paper, growth is taken to mean the production of more goods and services, using either more resources or the available resources more effectively. Growth is also interpreted as the availability of new or improved products, even if in absolute terms the number of goods remains the same.

33. The imitation lag is reduced as the number of competitors and the relative speed of market penetration increases. Imitation results in a redistribution of profits whenever imitators collude with innovators to restrict price reductions to consumers.

34. It is clear, however, that innovation does widen the quality gap and thus prolong the time prior to imitation by other firms. Innovation will also allow the firm to maintain its reputation as quality leader.

35. It makes little sense to have a policy that subsidizes imitation in an industry with two quality leaders. In industries where product technology is already shared, adding a new firm simply reduces the profits per imitator with little impact on economic growth.

36. The budget deficit, estimated at 15 to 20 percent of GNP (Hanson 1990), has been financed primarily by printing money, generating an estimated 25–30 percent growth rate in the money supply. Unofficial estimates put inflation at 20 percent in 1989–90 (Shmelev 1990).

Bibliography

Adams, Walter. 1961. *The Structure of American Industry*. 3d ed. New York: Macmillan.

Adams, Walter and Joel Dirlan. 1964. "Steel Imports and Vertical Oligopoly Power." *American Economic Review* (September): pp. 626–55.

Amann, R., and J. Cooper, eds. 1982. *Industrial Innovation in the Soviet Union.* New Haven: Yale University Press.

Amann, R., J. Cooper, and R. W. Davies, eds. 1977. *The Technological Level of Soviet Industry*. New Haven: Yale University Press.

Arakelyan, A. 1987. "Ways of Production Intensification." *Voprosy ekonomiki,* no. 3 (March): pp. 22–30.

Arbuzov, I. 1987. "Production Potential and Production Capabilities." *Planovoye khoziaistvo,* no. 2 (February): pp. 3–10.

Arrow, Kenneth. 1962. "Economic Welfare and the Allocation of Resources for Invention." In National Bureau of Economic Research, *The Rate and Direction of Inventive Activity*: pp. 619–25. Princeton: Princeton University Press.

Baldwin, William, and Gerald Childs. 1969. "The Fast Second and Rivalry in Research and Development." *Southern Economic Journal,* vol. 36, no. 1, pp. 18–24.

Baumol, William. 1990. "Entrepreneurship: Productive, Unproductive, Destructive." *Journal of Political Economy,* vol. 98, no. 5, pp. 893–921.

Bergson, Abram. 1978. "Managerial Risks and Rewards in Public Enterprise." *Journal of Comparative Economics,* vol. 2, no. 3, pp. 211–25.

———. 1983. "Technological Progress." In *The Soviet Economy: Towards the Year 2000,* ed. A. Bergson and H. Levine, pp. 34–78. New York: Allen and Unwin.

Berliner, Joseph. 1976. *The Innovation Decision in Soviet Industry*. Cambridge: MIT Press.

———. 1988. *Soviet Industry from Stalin to Gorbachev: Essays on Management and Innovation.* Ithaca, NY: Cornell University Press.

Bliakhman, L. S. 1968. "Nauka kak otrasl′ proizvodstvennoi deiatelnosti." In *Voprosy ekonomiki i planirovaniia nauchnykh issledovanii,* ed. L. S. Bliakhman. Leningrad: Izd. Leningradskogo universiteta.

Boretsky, M. 1966. "Comparative Progress in Technology, Productivity, and Economic Efficiency: USSR vs USA." In Joint Economic Committee, *New Directions in the Soviet Economy*. Washington, DC: United States Government Printing Office.

Brada, Josef. 1989. "The Comparative Economics of Bankruptcy: Bankruptcy in Socialist and Labor-Managed Economies." Department of Economics, Arizona State University. Mimeographed.

Brozen, Yale. 1951. "Invention, Innovation, and Imitation." *American Economic Review,* vol. 41 (Papers and Proceedings), pp. 239–57.

Cheng, Leonard. 1989. "Technological Innovation versus Imitation: A Decision-Theoretic Approach." University of Florida, Gainesville. Mimeographed.

Cohn, S. 1976. "Deficiencies in Soviet Investment Policies and the Technological Imperative." In Joint Economic Committee, *Soviet Economy in a New Perspective*. Washington, DC: United States Government Printing Office.

————. 1979. "Soviet Replacement Investment." In Joint Economic Committee, *Soviet Economy in a Time of Change*. Washington, DC: United States Government Printing Office.

Dasgupta, Partha. 1988. "Patents, Priority and Imitation or, the Economics of Races and Waiting Games." *Economic Journal*, vol. 98 (March): pp. 66–80.

Dorfman, Nancy. 1987. *Innovation and Market Structure: Lessons from the Computer and Semiconductor Industries*. Cambridge: Harper and Row.

Dubrovskii, K.I., and Yu Ekaterinoslavskii. 1976. *Upravlenie nauchno-tekhnicheskim razvitiem proizvodstvennykh ob"edinenie: Informatsionnyi aspekt*. Moscow: Ekonomika.

Freeman, Christopher. 1986. *The Economics of Industrial Innovation*. Cambridge: MIT Press.

Gavrilov, E. I. 1975. *Ekonomika i effektivnost' nauchno-tekhnicheskogo progressa*. Minsk: Vysheishaia shkola.

Gort, Michael, and Steven Klepper. 1982. "Time Paths in the Diffusion of Product Innovations." *Economic Journal*, vol. 92 (September): pp. 630–53.

Granick, David. 1967. *Soviet Metal-Fabricating and Economic Development: Practice versus Policy*. Madison: University of Wisconsin Press.

————. 1987. *Job Rights in the Soviet Union: Their Consequences*. New York and London: Cambridge University Press.

Green, D., and Herbert S. Levine. 1977. "Macroeconomic Evidence on the Value of Machinery Imports to the Soviet Union." In *Soviet Science and Technology: Domestic and Foreign Perspectives*, ed. J. Thomas and V. Kruse-Vaucienne. Washington, DC: George Washington University Press.

Griliches, Zvi. 1984. *R&D Patents and Productivity*. Chicago: University of Chicago Press.

————, ed. 1986. "Productivity, R&D, and Basic Research at the Firm Level in the 1970s." *American Economic Review* (March): pp. 141–54.

Griliches, Zvi, and F. R. Lichtenberg. 1984. "Interindustry Technology Flows and Productivity Growth: A Reexamination." *Review of Economics and Statistics*, vol. 66, no. 2 (May): pp. 324–29.

Grossman, Gene. 1989. "Quality Ladders in the Theory of Growth." Discussion Papers in Economics, Woodrow Wilson School, Princeton University.

Grossman, Gene, and E. Helpman. 1989. "Product Development and International Trade." *Journal of Political Economy*, vol. 90 (December): pp. 1261–1283.

Grossman, Gregory. 1960. "Soviet Growth: Routine, Inertia, and Pressure." *American Economic Review* (May): pp. 50–64.

————. 1966. "Innovation and Information in the Soviet Economy." *American Economic Review, vol. 56, no. 2* (May): pp. 118–29.

Gustafson, Thane. 1990. "Soviet Reforms: A Roundtable Discussion." American Association for the Advancement of Slavic Studies, Washington, DC (October).

Hanson, Philip. 1976. "International Technology Transfer from the West to the USSR." In Joint Economic Committee, *Soviet Economy in a Time of Change*. Washington, DC: United States Government Printing Office.

Hanson, Philip, and M. Hill. 1976. "Soviet Assimilation of Western Technology." In Joint Economic Committee, *Soviet Economy in a Time of Change*. Washington, DC: United States Government Printing Office.

Hayes, Robert, and William Abernathy. 1980. "Managing Our Way to Economic Decline." *Harvard Business Review*, vol. 58 (July-August): pp. 67–77.

Hill, Malcolm, and Richard McKay. 1988. *Soviet Product Quality*. New York: St. Martin's Press.

Jaffe, A. B. 1986. "Technological Opportunity and Spillovers of R&D: Evidence from Firms' Patents, Profits and Market Value." *American Economic Review*, vol. 76, pp. 984–1001.

Jensen, Richard. 1982. "Adoption and Diffusion of an Innovation of Uncertain Profitability." *Journal of Economic Theory*, vol. 27, pp. 182–93.

Jensen, Richard, and Marie Thursby. 1987. "A Decision Theoretic Model of Innovation, Technology Transfer, and Trade." *Review of Economic Studies*, vol. 54, pp. 631–47.

Jovanovic, Boyan, and Glenn MacDonald. 1988. "Competitive Diffusion." Department of Economics, New York University. Mimeographed.

Kamien, Morton, and Nancy Schwartz. 1982. *Market Structure and Innovation*. Cambridge: Cambridge University Press.

Katz, Michael, and Carl Shapiro. 1987. "R&D Rivalry with Licensing or Imitation." *American Economic Review*, vol. 77 (June): pp. 402–20.

Kushlin, V. I. 1976. *Uskorenie vnedreniia nauchnykh dostizhenii v proizvodstvo*. Moscow: Ekonomika.

Leary, Neil, and Judith Thornton. 1989. "Are Socialist Industries Inoculated against Innovation? A Case Study of Technological Change in Steelmaking." *Comparative Economic Studies*, vol. 31, no. 2 (Summer): pp. 42–65.

Levin, R. C., W. M. Cohen, and D. C. Mowrey. 1985. "R&D Appropriability, Opportunity and Market Structure: New Evidence on Some Schumpeterian Hypotheses." *American Economic Review*, vol. 75 (May): pp. 20–24.

Linz, Susan J. 1986. "Emigrants as Expert-Informants on Soviet Management Decision-Making." *Comparative Economic Studies* (Fall): pp. 64–89.

———. 1988. "Managerial Autonomy in Soviet Firms." *Soviet Studies* (April): pp. 175–95.

———. 1990. "The Soviet Economy in Transition: A Resurgence of Reform." Department of Economics, Michigan State University (May). Mimeographed.

Linz, Susan J., and Judith Thornton. 1988. "A Preliminary Analysis of the Demand for Innovation: Evidence from the Soviet Interview Project." Part I of the Final Report on the Science and Technology Project delivered to the National Council for Soviet and East European Research (December).

Loury, G. 1979. "Market Structure and Innovation." *Quarterly Journal of Economics* (August): pp. 395–410.

Lowenhardt, J. 1974. "The Tale of the Torch: Scientist-Entrepreneurs in the Soviet Union." *Survey*, vol. 20 (Autumn), pp. 113–21.

Mansfield, Edwin. 1963. "The Speed of Response of Firms to New Techniques." *Quarterly Journal of Economics*, vol. 67 (May): pp. 290–309.

———. 1968. *Industrial Research and Technological Innovation*. New York: Norton.

———. 1980. "Basic Research and Productivity Increase in Manufacturing." *American Economic Review*, vol. 70, no. 5 (December): pp. 863–74.

———. 1988. "Industrial R&D in Japan and the United States: A Comparative Study." *American Economic Review*, vol. 78 (May), Papers and Proceedings, pp. 223–28.

Mansfield, Edwin et al. 1982. *Technology Transfer, Productivity, and Economic Policy.* New York: W. W. Norton.

Mansfield, Edwin, Mark Schwartz, and Samuel Wagner. 1981. "Imitation Costs and Patents: An Empirical Study." *Economic Journal,* vol. 91, pp. 907–18.

Martens, J., and J. Young. 1979. "Soviet Implementation of Domestic Inventions." In Joint Economic Committee, *Soviet Economy in a New Perspective.* Washington, DC: United States Government Printing Office.

Nelson, Richard, and Sidney Winter. 1982. *An Evolutionary Theory of Economic Change.* Cambridge: Harvard University Press.

Nolting, L. 1973. *Sources of Financing the Stages of Research, Development, and Innovation Cycle in the USSR.* Washington, DC: United States Department of Commerce.

———. 1976. *Financing of Research, Development and Innovation in the USSR by Type of Performer.* Washington, DC: United States Department of Commerce.

Oshima, Keichi. 1973. "Research and Development and Economic Growth in Japan." In *Science and Technology in Economic Growth,* ed. B. R. Williams. New York: Wiley.

———. 1987. "The High Technology Gap: A View from Japan." In *A High Technology Gap? Europe, America, and Japan,* ed. Andrew Pierre, pp. 88–114. New York: Council on Foreign Relations.

Patrick, Hugh. 1986. "Japanese High Technology Industrial Policy in Comparative Context." In *Japan's High Technology Industries,* ed. Hugh Patrick. Seattle: University of Washington Press.

Reingaum, Jennifer. 1982. "A Dynamic Game of R&D: Patent Protection and Competitive Behavior." *Econometrica,* vol. 50 (May): pp. 671–88.

Sato, Ryuzo. 1988. "The Technology Game and Dynamic Comparative Advantage: An Application to U.S.-Japan Competition." In *International Competitiveness,* ed. Michael Spence and Heather Hazard, pp. 373–98. Cambridge: Ballinger.

Scherer, F. M. 1965. "Firm Size, Market Structure, Opportunity, and the Output of Patented Inventions." *American Economic Review,* vol. 55 (December): pp. 1097–1125.

———. 1967. "Research and Development Resource Allocation under Rivalry." *Quarterly Journal of Economics,* vol. 81 (August): pp. 359–94.

———. 1980. *Industrial Market Structure and Economic Performance.* Chicago: Rand McNally.

———. 1989. *Innovation and Growth: Schumpeterian Perspectives.* Cambridge: MIT Press.

Schmitz, James. 1989. "Imitation, Entrepreneurship, and Long-Run Growth." *Journal of Political Economy,* vol. 97, pp. 721–39.

Schmookler, J. 1966. *Invention and Economic Growth.* Cambridge: Harvard University Press.

Schroeder, Gertrude. 1979. "The Soviet Economy on a Treadmill of 'Reforms'." In Joint Economic Committee, *Soviet Economy in a Time of Change,* pp. 312–40. Washington, DC: United States Government Printing Office.

Schumpeter, J. 1934. *The Theory of Economic Development.* Cambridge: Harvard University Press.

Segerstrom, Paul. 1990. "Innovation, Imitation, and Economic Growth." Department of Economics, Michigan State University. Mimeographed.

Shrivastava, Paul, and William Souder. 1987. "The Strategic Management of Technological Innovations: A Review and a Model." *Journal of Management Studies,* vol. 24 (January): pp. 25–41.

Shteingauz, V. I. 1973. "Novye organizatsionnye formy sviazi nauki s pro-izvodstvom." In *Ekonomiki i organizatsiia promyshlennogo proizvodstva,* vol. 3, pp. 44–52. Novosibirsk: Nauka.

———. 1976. *Ekonomicheskie problemy realizatsii nauchno-tekhnicheskikh razrabotok.* Moscow: Nauka.

Shy, Oz. 1987. "Innovation Strategies and the Product Cycle." Department of Economics, State University of New York, Albany. Mimeographed.

Solow, Robert. 1957. "Technical Change and the Aggregate Production Function." *Review of Economics and Statistics* (August): pp. 312–20.

Spence, Michael. 1984. "Cost Reduction, Competition, and Industry Performance." *Econometrica,* vol. 52 (January): pp. 101–21.

Spencer, Barbara, and James Brander. 1983. "International R&D Rivalry and Industrial Strategy." *Review of Economic Studies,* vol. 50, pp. 707–22.

Sutton, Anthony. 1968–73. *Western Technology and Soviet Economic Development,* 3 vols. Stanford: Hoover Institution.

Tedstrom, John E., ed. 1990. *Socialism,* Perestroika *and the Dilemmas of Soviet Economic Reform.* Boulder, CO: Westview Press.

———. 1990a. "The Shatalin Plan and Industrial Conversion." *Report on the USSR* (Radio Liberty), vol. 12, no. 6. 16 November): pp. 8–10.

Telser, Lester. 1982. "A Theory of Innovation and Its Effects." *Bell Journal of Economics,* vol. 13 (Spring): pp. 69–92.

Thornton, Judith. 1985. "Prices and Technological Choice in Soviet Electric Power." Department of Economics, University of Washington. Mimeographed.

———. 1986. "The Technological Limits to Soviet Power." Department of Economics, University of Washington. Mimeographed.

———. 1988. "Barriers to Innovation: Evidence from Interviews." Department of Economics, University of Washington. Mimeographed.

———. 1990. "Crisis and Change in Soviet Information Technologies." Department of Economics, University of Washington. Mimeographed.

Thornton, Judith and Susan J. Linz. 1988. "The Supply of Innovation: The Relevance of Interview Evidence to *Perestroika.*" Part II of the final report on the Science and Technology Project delivered to the National Council for Soviet and East European Research (December).

Toda, Y. 1979. "Technology Transfer to the USSR." *Journal of Comparative Economics,* vol. 3, pp. 181–94.

Weitzman, Martin. 1979. "Technology Transfer to the USSR." *Journal of Comparative Economics,* vol. 3, pp. 167–77.

Zaychenko, A. 1988. "Risk and Independence of Innovative Activity." *Voprosy ekonomiki,* translated in JPRS-UEA–88–008 (May 12): pp. 27–33.

Ziegler, Charles. 1985. "Innovation and the Imitative Entrepreneur." *Journal of Economic Behavior and Organization,* vol. 6, pp. 103–21.

Science and Technology in Eastern Europe after the Flood: Rejoining the World

Steven W. Popper

Introduction

The flood metaphor describes the situation the science and technology practitioners of Eastern Europe now find themselves in. The economic, political, institutional, and cultural foundations upon which their working environment was predicated have been all but swept away. Especially for the East Europeans, who unlike their Soviet colleagues cannot entertain any illusions about their countries' R&D systems being sustainable as more or less independent entities, the next years will mean a startling period of transition. The transition will be one not only of domestic institutions but of fundamental orientation as well. East European science and technology[1] needs to rejoin the global mainstream to survive and to be effective. This paper discusses the means for doing so and the problems to be faced.

Students of Soviet and East European affairs are occasionally guilty of parochialism. This may also be said of analysts of the region's science and technology systems. Phenomena appearing to be satisfactorily explained by the peculiarities of orthodox Soviet-type institutions are sometimes not taken beyond the context of the local situation. Certainly, if the temptation to view all behavior as a result of system-specific institutions was great before, it receives reinforcement now as the systemic character of the general crisis becomes more stark. Yet, if the forces affecting the science and technology establishments in these countries are not viewed as part of larger, worldwide phenomena, some of the subtleties may be missed.

An example of this is the interpretation of the CMEA (Council for

Mutual Economic Assistance) Comprehensive Program (hereafter, "the Program") for Science and Technology to the Year 2000, signed by the heads of government in Moscow, December 1985. Its existence and origin would seem to be adequately explained by the autarkic nature of the economic systems prevailing in each member state, the closed and hierarchical character of national science establishments, and the previous record of CMEA in attempting to erect monstrosities of planning and coordination in a wide range of fields. Yet, at the same time, it cannot be entirely coincidental that one of the emergent themes in contemporary research, both basic and applied, throughout the world is increasing multilateralization and internationalization of research efforts in a wide range of fields. To fully appreciate the changes required of the East European science and technology systems to meet the challenges of the post-Communist future, the systems must not be viewed solely as reacting to the malignancies of what has gone before on the local scene. They need to be analyzed also for their viability as mechanisms for meeting the challenges being posed to science and technology organizations everywhere. In this light, this paper will consider the necessary changes needed to preserve the scientific tradition in Eastern Europe in the face of the recent upheavals and to implement rejoining, in the fullest sense, the mainstream of the world scientific community.

A Wider View of Contemporary Science

Although generalizations must be used with caution, several trends in the management and application of science are apparent across national frontiers and across systems. One appears to be the shift from viewing science as the "endless frontier," with ever-expanding scope and horizons, to seeing it as a constrained, steady state system forced to confront stringencies in funding allocations and limits to human and material resource endowments.[2] While one may certainly argue about the applicability of this image to individual countries, the general environment surrounding scientific pursuits often has a different feel than was the case in the fifties and sixties. Concern over government deficits and the accompanying need to trim subsidies by cutting back or placing former recipients of state subventions onto a self-financing basis are not solely Soviet or East European phenomena. This school of thought would argue they are manifestations of a worldwide, perhaps irreversible, structural change in the management, organization, and performance of science (Ziman 1987).

Coupled with this shift is a second major global trend: making science pay its way. Pressure is being placed on the science system externally by heightened expectations for the potential contribution from the science and technology base to national welfare. The code words are "science policy" and even "strategic science," but the real issue is international competitiveness reflecting a near universal concern for preserving or securing market share in several high-technology areas in the face of formidable pressures for reduction.[3] Science itself, then, is not the real object. Rather, these policies are driven in part by a paradigm viewing science as the progenitor of technological advance and the mainspring of the technology system. This theme and its problematic nature will be explored below in a slightly different context. The central tenets of strategic science are recapitulated on all continents, in both developed and developing countries, in technological leaders as well as among traditional technology followers.

A third trend stems partially from the first two. In an era where there is at least a perception of growing resource constraint and where, at the same time, great expectations are placed upon even fundamental basic research, scientific and technological pursuits are becoming more multilateral and international in several disparate fields.[4] Pooling of resources, especially in big-ticket scientific fields, would seem prudent. Further, becoming a member of a consortium appears to be a method of writing a partial insurance policy if competitiveness based upon any ensuing applications is a concern. A consortium member gambles its opportunity for gaining exclusive rights over any useful discoveries against the greater likelihood of being assured some slice of whatever issues from the cooperation.[5] Beyond these concerns, there is also a growing awareness that the public, affected by the public-good aspect of scientific knowledge, is more than just the domestic population of the country where original research is performed. Further, much of the impact, for good or for ill, necessarily carries across national frontiers. If any progress is to be made in several of the more pressing areas of concern addressed by science today, the effort must, as a matter of course, be international since the potential solutions are almost certain to be so.

It remains to be seen whether this trend is a harbinger of the future or a passing fancy that will prove incapable of handling the strains brought on by conflicting incentives inherent in any cooperation, especially as one draws closer to the applied end of the research continuum. Moreover, a considerable impetus comes from possibly transitory national political considerations, such as a desire to signal a willingness for closer cooperation among Europeans. For the present, the tendency is toward greater internationalization. In this respect, the CMEA program is less an aberrant attempt to intrude late Brezhnevite ideals of socialist internationalism into the world of scientific research and more of a piece with global trends.[6] In fact, four of the program's five main research areas—microelectronics, automation, new materials, and biotechnology—are recapitulated in virtually every other multifaceted international-research effort, among them EUREKA, ESPRIT, BRITE, RACE, CERN, ESA, and even SDI.[7]

The organizational arrangements for the modern science regime are still fluid and require greater definition. Decisions based upon current policies will crystallize them. If there truly has been a secular shift in the relations between established science and larger society, there is probably no way back to the traditional arrangement; but there are as yet many different ways forward. What is needed in such a system is a way to provide accountability, efficiency, excellence in performance, exploitable outputs, and adaptability to change while preserving individual creativity, time for ideas to mature, openness to criticism, hospitality to innovation, and respect for specialized expertise. The shifts required in Eastern Europe are no less and the need at least as great. They may be viewed as a special, if acute, case—if not the quintessence of the shift occurring on a global scale.

The Deluge

The challenge facing Western science establishments is how to conserve the cherished values enumerated at the end of the last section in the face of a changing environment. The problem facing the East Europeans is how to deal with the same set of external pressures and demands while, at the same time, somehow rebuilding many of the values they once held in common with other scientific communities, destroying the institutions of forty years in the process. The problem of transition in Eastern Europe is therefore twofold, involving a process of creative destruction on a scale unmatched elsewhere. It has greater urgency because the continued existence of many scientific establishments in the region may by no means be taken for granted.

The structure of the science and technology establishments of Eastern Europe will be familiar to students of Soviet R&D and needs no detailed elaboration here. A major defining element is the Academy of Sciences system. Fundamental research is performed in academy institutes and not in the universities, whose principal responsibility is teaching. Applied research, in turn, is largely conducted within the parallel system of ministerial R&D institutes organized by industry. This design assumes the presence of a well-informed central authority actively guiding research efforts to maximize the resources available to society. In practice, the result is more often considerable fragmentation of effort because of compartmentalization, so that even applied work quite often does not address the real needs of the clients.

Another characteristic was, of course, the political and ideological intrusion into personnel- and priority-allocation decisions present in all countries of Eastern Europe to varying degrees. This was reinforced by the nature of financial support of science and technology. Although principles of self-finance had entered into the R&D systems of several countries, for most practical purposes institute working funds were allocated from the central budget.

A further legacy of the past is the insufficiency of information available to practitioners, particularly at the applied end of the R&D spectrum. Lack of meaningful price information compounded by internal accounting systems not suited to the needs of institutes and enterprises and combined with contrived systems of incentives has led to poor technology choice and inefficient utilization of R&D resources. This aspect of the former system has been a greater factor in denying Eastern Europe the practical fruits of technology than was COCOM (the Coordinating Committee on Multilateral Export Control).

Problems of Transition

The system as it has existed will not continue to serve the needs of East European science and technology establishments. Many factors will be forcing change.

Economic

Clearly, as with other East European institutions, many of the greatest problems will stem from the general economic crisis coupled with the need to transform the economic systems to market-based ones more responsive to present needs.

Among other things, this will mean a funding crisis. As part of a macroeconomic-stabilization package, each country will be reducing central expenditures from which resources for fundamental and applied research have almost exclusively come. This break will be even more sharply felt because, in the past, the fortunes of basic science have corresponded with the fortunes of defense-establishment budgets. There is no question about the plunge these will take in the short to medium term.[8]

Science establishments will be asked to become self-sustaining inasmuch as is practical (and perhaps even more than that). But here will lie the rub. While the need for economic stabilization is clear and pressing, the transformation of the economic apparatus to a new form is likely to take time and is even sometimes viewed as a later stage of the process.[9] In fact it is integral. If scientific institutions are forced to rely upon their own resources, there must be an external and internal environment making this possible. In large measure this milieu does not exist. Both inside research institutions and in industry there remain fundamental problems of a systemic character in supporting risk-taking behavior by researchers, developers, and would-be innovators. These will not be resolved before there is a profound shift in the nature of the information available to decision makers (i.e., a price system reflecting true scarcity values and real resource costs) and a change in the incentive structure confronting them. This last, in turn, will rest squarely upon a radical redefinition of ownership rights, likely to be the most contentious and knotty problem faced during economic transition. Yet, it is the source of the chronic, systemic failure to create applications from research findings. This will not change without a complete redrafting of the system within which the relevant decisions are made.

Too drastic an application of the tenet of self-financing of research could also adversely affect the research programs of East European science. There is danger of forcing too short-term an approach on research, thereby reducing more fundamental and theoretical pursuits while increasing the potential cost of attempting major leaps. Whether this outcome might not be desirable is a question of policy to be treated below.

Political

The revolutions in domestic politics are also reflected in the world of science. In particular, the general crisis of authority caused by changing concepts of legitimacy ushered in by the revolutionary changes of 1989 has left serious rifts within the microcosm of the academy system and its research organs. Strains have developed between the young research workers who have imbibed deeply the spirit of '89 and their section heads and institute directors who are viewed as often having gained their place through obsequiousness and political rectitude rather than scientific aptitude or managerial ability.[10] The attempts of the latter to trim their sails quickly to catch the new wind and maintain their positions have led to considerable preoccupation with internal politics at the expense of scientific endeavors. There is a need to normalize the situation by bringing the legitimacy of institute leadership into accord with the prevailing notions of pluralism, democratic process, and merit. At the same time, there is only so much disruption existing institutions can take. This tension between change and preservation will be one of the strongest undercurrents in the political and social life of East European research institutions in the near term.

The internal crisis of authority may well be matched by a challenge to the authority of science within the society as a whole. Again, a tendency toward the abuse of science and its practitioners, after having placed them on the pedestal of official civic (socialist) virtue for so long, would be a local ripple of a larger current taking place outside Eastern Europe as well (Teich 1990). In the West, this abuse is not fundamental. Science is called to task on issues of animal rights, falsification of data and research, and participation in weapons research, to name a few specific cases. It still retains some blanket protection to its authority because it is largely viewed as being apolitical and objective. In Eastern Europe, however, where one of the single-party regimes' claims on the right to rule was their ability to accelerate the pace of progress by bringing science to bear in a planned and coordinated manner on the problems attending the creation of national wealth, scientists may not find themselves so insulated. Science and the scientific establishment were closely identified with the old Communist order in each country of the region. This may well cause a falloff in the prestige of science in the short term, especially as the full range of problems left strewn in the wreckage of the old regimes becomes more clear to the public.[11] The practical effect might be large if it leads to pressure to close facilities or to even more draconian cuts in central-budget allocations already due for considerable slashing.[12]

The cadre problems manifested in the phenomenon of brain drain partially reflect the discrediting of the existing scientific establishment and partially the general loss of prestige of science, but they cut across several other themes as well. The ability of East European establishments to retain and effectively utilize trained personnel is in question. This theme will be treated in a wider context below.

The Crisis of Institutions

The economic and political problems of East European science will provide the background for the major task of reformulating or rebuilding the institutions that are to govern, direct, and judge the results of scientific research in each country. Many would wish for stability in the political and economic environment so this business could proceed in a deliberate and considered fashion, rather than be forced in an atmosphere of crisis. Yet, it may well be that the revolutionary temper of the times may allow and, indeed, may force a more profound reorganization than might otherwise be the case.

The questions to be answered are truly fundamental. How does one encourage autonomous decision making in such formerly centralized systems? Basic approaches to funding, project choice, technology assessment, and assignment of priority will need redrafting as the hierarchies governing these decisions are replaced.

Especially during the era of diminished resources Eastern Europe is now entering, there must be decision making, using yet-to-be-specified means, to assess where assets are best allocated, while at the same time introducing new flows of information from lower-level units. Yet, Eastern Europe carries the legacy of the past: concepts and mechanisms for peer review, both *ex post* in assessment and *ex ante* in project choice, have either never existed or have lain dormant for forty years and more. Networks of contacts existed and continue to exist, but these often are based more on political calculation than acknowledgment of expertise and are coming under increasing pressure as change sweeps these societies.

Priority will need to be assigned not only to individual projects but to entire institutes as well.[13] In doing so, the East Europeans may suffer from a lack of direct experience with other models of organizing and sustaining R&D work, and training because of travel and other restrictions in place for the entire working lives of virtually all current practitioners.

There may also be a more subtle institutional handicap. The intellectual and institutional architecture of the Soviet-type economies' *science and technology* system, taken as a whole, and in particular the perception of how science relates to technology, may perhaps be particularly ill-adapted to the process of transition.

The point may be illustrated by postulating the existence of two identifiably different systems for defining the relationship between science and technology. The first analogizes from the apparent experience of the Second World War where science-derived weapons such as radar, sonar, and the atomic bomb, to name but a few, were the source of technological supremacy and ultimate victory. This view, in its extreme, would hold technology to be an appendage of a system driven by basic science. Process is linear: fundamental findings about natural law lead to applied findings, which are then taken by technologists to be developed into specific applications. Implicitly, this creates a *de facto* caste system. High *pukka* Science pushes back the frontiers defining the limits of human knowledge while lower caste Technology follows in its course, making use of what is left in the wake of pathbreaking research.

There is, of course, a good deal of caricature in this exposition. Yet, it does capture the essentials of a philosophical system prevailing in many venues. In part, it provides the intellectual foundation behind calls to "make science pay" by marshaling strategic-science policy to contest for economic competitiveness. Such a view argues that the scientific revolution has finally, in this century, come into its own. However, in positing such linearity, this paradigm may overstate the dominant role of science and minimize the complex relationship between it and technology.

We may identify a second paradigm, opposed to the linear, science-driven one, which does not make the scientific revolution as central an event. Rather, it would recognize science (the path of accumulating knowledge by organized conjecture) and technology (the path of accumulating knowledge by making) as two discrete, if highly integrated, avenues of human progress, each with a long-shared history, but with technology possessing the longer independent existence. In this view it is less certain which is the tail and which the dog. In fact, an extreme position would relegate science to the position of exploring the new worlds that technological developments in machine making and instrumentation have disclosed.[14] It is certainly quite often the case that technology enables science by presenting it with new problems, on the one hand, and new tools for observation on the other. A science and technology system built upon this latter view would be less caste-oriented, and organizations would be designed to admit the possibility of useful notions being generated in either sector to then be jointly developed by input from both.[15]

The suggestion to be derived from these two postulated approaches is that it is possible for a science and technology system organized according to the first set of principles to generate an impressive store of knowledge and Nobel prizes, while gradually losing market share and competitive position to other systems admitting of more complicated feedback relations between two more or less free standing approaches to mastering nature. For the present purpose, a hypothesis may then be derived. The system of Soviet science and technology, and by transplantation that of the East Europeans, may be more accurately modeled by the first set of postulates than by the second. This may have been partially conscious with the Soviets seeking to follow in the path of those nations that the Hegelian process of history had determined to be winners.[16] In part, this paradigm may have come by default as a derivative of the Soviet system's first principles for general economic and social organization. In a strictly hierarchical system of centralized control, where power is held by one group because of their claim to an uniquely acute understanding of history, there would be a natural gravitation to a philosophical construct that provided a simple, unidirectional model for technological interaction. The planned economy bespeaks commitment to a linear view of process. The fact that in some East European countries, as will be explored below, science could be identified with the coming to power of the Communist regimes would strengthen this predisposition.

Is it accurate to characterize the science and technology systems of Eastern Europe in this manner? After all, the dual existence of both academy and industrial ministry R&D institutes would seem to pay court to the notion of two free standing structures for technology development. And wasn't there always a strong predisposition on the part of the political leadership to bring science to the service of the economy by supporting technological development, even to the detriment of long-standing traditions of excellence in basic research? Definitive answers can only be forthcoming after further study. However, the distinction between the two systems developed above does not lie solely or even largely in the difference between a unitary linear system and one where there formally exist two parallel hierarchies. Rather, the difference would be characterized by the degree of complexity permitted in the interactions between the structure of science and that of technology; between linear interaction along a well-defined time course and interaction that is more frequent, less well-scheduled, and where the direction of flow and influence is much more complex than the linear model would suggest. Here, the near hermetic closure of the dual academy and ministerial systems of research and the near inadmissibility of either being equally likely to have a profound influence on the other would mean a linear type of interaction would be the best that could be hoped for. Anything more complex could not be supported by the existing structure. The essence of the more complex version would be realized not in two hermetically closed structures, but rather in two sets of institutions, freestanding in some respects yet each vitally dependent on the other in ways too complex to adhere to any simple formulation of precedence. Beyond this, the very existence of an identifiable political predisposition to make science the driver for technological development provides a second marker.

The suggestion is not that this approach is wrong in every circumstance or even in fundamental concept. Rather, for the purpose of supporting technological development, especially in an era of profound transition such as Eastern Europe is now experiencing, it provides a more restrictive set of possibilities for organization and interaction and so may be more likely to prove nonadaptive in certain fields or environments.

How Does East European Science and Technology Differ?

Much, if not all, of what has been said above could apply to the Soviet case as well. This is certainly not surprising. Even though Eastern Europe was always too diverse a region to be accurately spoken of in the general terms usually dictated by expediency, in few spheres was the Soviet model so faithfully recapitulated than in the area of organization and administration of science. Are there, then, identifiable differences affecting the qualitative aspects of the transition problem?

There are two obvious areas of difference. The first is that the East Europeans would appear to fall into two classes. There are those who, like the Soviet Union, have an independent scientific tradition of some long standing (East Germany, Poland, and Hungary are certainly among them) and those like Bulgaria and Romania whose scientific tradition stretches back less far and whose domestic scientific communities are more a development of, or are contemporaneous with, the Communist era. While such a generalization is heroic in its scope, this factor may prove to be of more than historical interest. The distinction is not intended to impugn the native genius of the Bulgarian or Romanian peoples nor to disparage their ability to produce scientists of great stature. Rather, it suggests there may be greater difficulty in generating rank-and-file scientific workers who, nevertheless, are able to make a contribution of significance in global terms.

Further, to the extent that the others possess a scientific tradition more closely bound to the West, it may be easier to make the transition back to what for them will be an older system of organization, but one with which they can identify and for which they retain at least institutional, if not personal, memory. The transition may prove more difficult in countries where science traditions are more the creature of the Communist system.

Connected with this, while their Soviet colleagues have had only attenuated contact with the West (surely since 1928, and perhaps since 1914–17 in the main), East European scientists have been split off from the mainstream of the scientific community only since the period 1939–47. The additional quarter century or more of contact may make a difference in the same way more recent experience with functioning markets appears to make a difference in the pace and popular acceptance of profound reforms in the economic sphere.

The other major difference, of course, is in the size of national science establishments. In spite of the impressive official statistics on percentage of population with higher degrees and on the number of scientific workers in the establishments of Eastern Europe, there is a vast difference in absolute size between these establishments and the self-contained universe of the Soviet scientific and R&D hierarchies. The quantitative difference may be so great as to constitute a qualitative one. To state it as a hypothesis: as with much else in the orthodox Soviet model for economic development and social organization, what might exist as a cumbersome, perhaps inefficient, but ultimately workable system for the Soviet Union has a disproportionately ill effect on smaller countries. While autarky in economic development may have been a viable policy option for the Soviet Union, it was a disaster in the countries of Eastern Europe where it was adhered to. Similarly, perhaps maintaining a largely self-contained and self-sufficient scientific-research facility is viable, with many qualifications, for the Soviet Union, while for the smaller East European establishments the relative lack of contact with other scientific communities leaves them enfeebled and/or overly dependent on the Soviet Union.

The shortfalls in both areas have been ameliorated by international contact and commerce through the instrument of CMEA. At least since 1971 and the drafting of CMEA's Complex Program, the desirability of more intense scientific interaction between member states has been an increasingly central concern of CMEA. The early attempts, however, showed little result, with a few exceptions.[17] Efforts at integration of scientific work and drawing together R&D communities reached their apogee in the Comprehensive Program of 1985 mentioned above. Below the five major directions for cooperation were listed 93 main tasks, divided further into no less than 629 specific projects.

What set the 1985 program apart from its predecessors, according to the Soviets,[18] was the interconnection between the various research tasks, the emphasis on putting results into actual production, and the unifying concept of "direct ties" between Soviet and East European scientific and R&D establishments.

The program is too large a topic to discuss usefully here.[19] For the present purpose, it is interesting to note that the response to the program varied among East Europeans. The fundamental concerns included a perceived danger of increased dependence upon the Soviet scientific and research hierarchy as well as skepticism over whether the results were likely to justify the costs to the participants. All of the head organizations for the 93 tasks were Soviet entities. Moreover, in spite of perceived mutual benefits ensuing from the cooperation, there existed potential Soviet unilateral benefits: gearing CMEA high-technology output to best suit Soviet needs through the setting of standards; placing Soviet organizations in a better position to control East European R&D output and more effectively monitor the quality of intra-CMEA trade; and controlling, if not actively restricting, scientific and technology contacts with the West.

Bulgaria and Czechoslovakia signed on readily to the program, the latter at some cost because of its already close economic and ideological dependence upon the Soviet Union. The Poles, who because of their economic travails were permitted to run a series of large deficits with the Soviets, were not in a position to be outspoken in their concern. Romania's Ceausescu was not so reticent and stated his opposition in vociferous terms, although this may have stemmed in part from having had a previous understanding with Chernenko on a Soviet raw-material delivery *quid pro quo* going awry and not being made part of the final package.

The reticence of East Germany (GDR) and Hungary is most interesting. The East Germans believed the differential between the qualitative level of their scientific and R&D cadre and that of the Soviets would work to the detriment of their technological development. There might also have been present a fear that the special relationship with West Germany, increasingly important to the GDR during the eighties, might be jeopardized by too eager an acceptance of closer intra-CMEA cooperation. The Hungarian objection was more subtle but seems to have derived largely from the damage that might be done to two decades of Hungarian efforts to remerge their scientific community with the larger international, and particularly Western, mainstream.

Whatever the ultimate effect of the program might have been, the current moribund state of CMEA and the manifold questions about its future in any form have put implementation on permanent hold. The nature of cooperation and interaction in any form between scientists in the member states becomes a matter for renegotiation.[20] As these ties are sundered, so also are most present arrangements for multilateralism in East European research efforts. Some science areas (nuclear research, for example), even in countries like Hungary with an exceptional Western orientation, are heavily dependent upon contact and exchange with Soviet institutions and scholars. There is potential for creating a large void, not only in areas of collegial interaction, joint activities, and data exchange but also for access to research facilities and equipment not available in each East European country.

All of the research institutions of the former CMEA states will now be searching for new partners for research and alternative means of support and access to the global scientific community.[21] This may, then, provide the greatest difference between the fate of science in the Soviet Union and in Eastern Europe. On the whole the East Europeans dispose of fewer resources to attract potential science and technology partners. While for Soviet science, serious readjustment or even major retrenchments may be necessary, in Eastern Europe entire scientific disciplines may be on the brink of extinction (or at least a savage pounding), because given their relatively small scale they may be unable to survive the limited access to financial and political resources looming in the near term.

Transition: Solutions and Pitfalls

The quest for a transition path toward a new system for supporting national scientific and technological development may be broken into two parts. The first step is to ask what each nation may require from its science and technology establishment to effect the larger transformation of the domestic economy. This begs the question of the time for transition and how soon positive results will be required. Putting this aside for the moment, there still remain substantial questions about the relationship between science and technology, and their relationship, separately and as a system, to economic development and international competitiveness. The answers to these questions have by no means been satisfactorily resolved in the relatively more stable world of the West.

It becomes particularly difficult during a period of such tumultuous change to ask what the positive role of science and technology may be in the economic transformation of the countries of Eastern Europe. Any enumeration of the input side of the ledger would show that the East European countries possess considerable human and capital resources in both science and technology development. Establishments are well-endowed with personnel and, in some instances, equipment (R&D expenditure taking a proportionately larger share of national income than is usual in the West), but are peripheral as far as the world scientific community is concerned.[22] Their inputs have failed to generate wealth-creating outputs because of a systemic inability to use resources, especially information resources, effectively. For the countries lacking earlier scientific traditions, the past forty years have combined the phenomena of catch-up with relative isolation from other scientific communities except for the integration efforts brokered through CMEA.

These factors would seem to place the countries of Eastern Europe in the classic position of technological followers rather than innovators. As a matter of policy, then, would it be better to accept this situation or to try to achieve a breakout into the ranks of the technology leaders in some areas of comparative advantage? Various entry costs would impose considerable obstacles on the latter course. These countries might appear better placed to profit from the diffusion of innovations by taking up the classic position of product-cycle followers, assuming their economic houses can be put into order. Yet, to follow this course raises a specter of permanent dependency; it also calls into question the ability of the currently existing industrial structure to assimilate technology quickly and efficiently. These countries have suffered nowhere near so much from the technology embargo imposed by the West through COCOM as from their own systemic inability to elicit from capital equipment the fullest measure of capability embodied in the technology they do import (Popper 1990).

Trying to capture a technological lead, because of resource limitations, will entail somehow identifying and generating winners. The experience of the past, both in the former Soviet bloc as well as in the West, has shown the difficulty of this strategy. Two main approaches may be identified. The first route would be the strategic-management ("science policy") approach implicit in the "science leads" paradigm of science and technology. This would come quite naturally to East European policymakers since it requires choices to be made and implemented by a technology planning staff.[23] This would be a seductively dangerous course for any government to follow but especially counterproductive in an Eastern Europe desperately in need of fundamental change. Given the current situation, any attempts to move in this direction in Eastern Europe are likely to be overly hierarchical, to be laboring under the paradigmatic burden described in the sections above, and ultimately prone to making the wrong guesses.

The problem with strategies based on the linear integrated model, even if one believes it captures the essence of science's role, is that one cannot know *ex ante* what bets will pay off and, more naggingly, when. In its essence, the strategic-management route attempts to deal with the uncertainties inherent in technology development by imposing a structure on the future. Rarely do such efforts prove effective; the future seldom pays court to the exigencies of today. A multi pronged approach to future technology assessment and development is more likely to allow more of one's bets to remain covered.

The alternative, then, is to employ a less centralized, more opportunistic market-oriented strategy by maximizing the number of development centers to increase the likelihood of coming up with winning combinations. But this, again, is a game the East Europeans are currently less well suited to play than are (potentially) the Soviets. It is precisely the relative paucity of alternative centers that distinguishes the East European technology-development base. Even if this were not so, pursuit of this approach requires the existence of domestic consumers who are authoritative, both in the sense of being knowledgeable and sovereign over purchase decisions, to make international success even remotely likely. In other words, choices must still be made over what avenues would be fruitful to exploit. There are no mechanisms inherent in the science and technology institutions of Eastern Europe today, nor in the wider economy, to make these choices in an informed way. Once again, the connection between wealth-generating technological development and the *sine qua non* of profound economic reform appears ineluctable.

The "science on the market" approach also carries a potential cost, depending on how sweepingly it is applied to existing assets. In institutes placed on a self-financing or *polnii khozraschet* basis in the Soviet Union and Eastern Europe there have been complaints about the qualitative changes this status brings to research programs (see Panova and Matveev 1989). The initial response has been to be more result-oriented and to adopt a more short-term planning horizon. However, this reduces the amount of basic and, it is claimed, potentially path-breaking, but necessarily more risky, applied research.[24]

However this balance is resolved in the long term, in the foreseeable future the technology component of the science and technology system may receive most emphasis and priority. This is perhaps as it ought to be. The "science leads" technology-development path, bespeaking a need for a large force in basic research, is not the only way to proceed. Yet, a drastic reduction in basic-science funding would entail considerable cost in human terms and run the risk of frustrating the hopes of the generation of young researchers whose aspirations played a large part in bringing about the revolutions of 1989. Further, the fundamental problems of how to decide priority and how to determine what assets to let go remain.

This, then, raises the second major question surrounding the process of transition, namely, what the scientific and R&D establishments in Eastern Europe need from their respective nations in order to prosper. Clearly, the primary issue will be funding in milieus where the former financial arrangements have been, or are likely to be, completely overturned. Domestic resources alone are not likely to prove sufficient to maintain current establishments even after considerable transformation. All signs point in the short term to a need for greater research cooperation, joint ventures with foreign partners, and participation in international consortia. This applies equally to basic and applied research, but may be more crucial to the survival of the former.

Western involvement in East European science and technology has the potential for resolving more than just the financial crunch. Participation by Western governments and private commercial interests in the science and technology systems of Eastern Europe could prove crucial in helping to determine where priority should be set for R&D activity, which assets to develop and which to forego, how to orient applied-research establishments toward the market, and how to fund basic research. This is perhaps the only avenue in the near term to achieve meaningful participation in multilateral endeavors and to become a part of the international flow of products and ideas.

In determining the actual form of cooperative assistance, it behooves both the East Europeans and their potential foreign partners, both sovereign and commercial, to treat the science and the technology components of the system more as separable, and less as antecedent and successor activities, in accord with the paradigm outlined above. This is not only because of the different nature of activity in each area but also because such a distinction allows the exact role for each potential Western player to be made more clear. As a side benefit, this would go far toward reconstituting the institutions of Eastern Europe in a direction probably more favorable to efficient use of R&D assets, but a direction toward which, as has been suggested above, they are not otherwise historically or philosophically disposed.

Preservation of the *basic* scientific-research base is important to a modern economy. Even if the second, coequal paradigm of science and technology is accepted, this does not relegate theoretical research to the same plane as opera—something the state should support because of its aesthetic and character-building values. The real contribution of public R&D spending is not so much the actual fruits of research as the skill building that occurs as part of the process. This process of training in the sciences is a necessary support for a higher level of technology activities (Pavitt 1988). This, again, suggests strategic science may be too myopic to make education a primary focus and so prove debilitating in the long run.[25] This is also a danger when basic science is excessively caught up in commercial competitive issues.

"Big science" projects and disciplines would not seem to loom large in the immediate future of East European scientific establishments, and this is perhaps as it should be. Yet, it is entirely possible that even what is worthy in these establishments may not be preserved over the short term without cooperative arrangements with Western governments and institutions. Cooperation may take many forms, ranging from exchanges and fellowships to actual cooperative research agreements between partner laboratories or institutes.[26] Contacts with, and Western assessments of, individual labs and workers will become crucial. In effect, an external selection factor will play a large part in helping local governments determine which of their assets are worthy of support and exploitation. In an era when Western governments are searching for ways to support Eastern Europe without compromising the pressures forcing change, this may prove one of the most fruitful areas for consideration.[27]

A similar connection with the West might prove crucial as a means for East European *applied* R&D personnel and institutes to escape the binds they find themselves in. But here the principal instruments should be joint ventures with and direct investment by Western commercial interests, not governments. These have the potential of providing vital funding resources while obviating the problems caused by fungibility and the indiscriminate targeting characteristic of other means of resource transfer. Formal contracts and agreements with foreign businesses familiar with the management and maintenance of effective research facilities would also go quite far in providing a tie-in to multilateral research efforts and easing the process of rejoining the international research community as fully participating members.

The Western partner would provide more than just money in return for use of a country's research assets. Lack of well-developed domestic markets, decades of enforced isolation, inexperience with techniques for management and priority choice consistent with the production of technologies suited to the needs of customers, and the institutional legacy left by an overtly ideological orientation to development leave the East European R&D establishments distinctly unprepared to make the choices facing them. They are underequipped to compete in the game of identifying and developing winners. The Western commercial partners can fulfill many of these functions. For years the East Europeans have been able to develop commercially useful technologies without the ability to recognize or exploit them.[28] The Western partner will provide the marketing- and technology-assessment infrastructure to fill the gap. In the course of this process, disembodied technology for management will be transferred.[29] The potential exists for creating new training systems: for restructuring the science management infrastructure, the policy-setting processes (peer review, etc.), research management, and marketing; for modernizing and retrofitting those branches of industry worth saving (and perhaps more important, tacitly indicating those areas it would be best to scrap); and for generally demonstrating how applied research can be used to make traditional industries competitive.

The extent of Western involvement and the outcomes likely to ensue are, at this writing, highly speculative. Much depends upon the attitudes of the East Europeans and their true willingness to change. The interest is at least present in the West and has been demonstrated by the number of ventures entered into already, the even greater number of firms expressing interest in more substantial efforts, and in government support through such measures as the SEED Act. This all need not necessarily come to pass, however. The East Europeans may find this approach too costly in domestic political terms to entertain. If this helping hand is not firmly grasped, however, there may be little chance of saving domestic science solely through local means; these countries may then find it impossible to truly live up to the technological potential they have, over the years, sacrificed so much to build.

Notes

1. The phrase "science and technology" will be used in this paper to connote a unitary system for generating the familiar outputs of research results, both basic and applied, and technological development. Hence the use of the singular verb. By Eastern Europe, I refer to the five states remaining from the former CMEA Six: Bulgaria, Czechoslovakia, Hungary, Poland, and Romania. The situation of the former German Democratic Republic is a special case.

2. The steady-state theme is owing to John Ziman (1987). Whether a general "limits to growth" model of science is universally applicable is open to dispute. The latter would appear not to allow for changes in forms and institutions and it abstracts the science system from the background of the larger systems within which it is contained and which it might also be transforming. Yet, the image certainly seems to capture many of the strains afflicting science today.

3. The phenomenon of shrinking market shares may be more statistical than economic in character: a larger number of technologically competent exporters

necessarily means a smaller proportional slice for almost all players.

4. In the ten years between 1976 and 1986, the fraction of scientific papers by French, West German, U.S., and Japanese scientists that were internationally co-authored at least doubled (Perry 1990.)

5. This phenomenon is not found exclusively in the domain of public efforts. In the private sphere, there is also a trend to bring smaller pieces together, to form cooperative industrial R&D efforts. The growing perception (almost certainly erroneous) is that competition on the international level may be more important than domestic competition. This has led to pressure on long-held policies designed to promote pluralism and decentralization.

6. This is not to suggest all members of CMEA were equally eager to join the program. On this point see Popper (1991).

7. The fifth area, nuclear energy, was out of fashion in the West at the time many of the more recent consortia were formed, but it was certainly among the earliest areas of cooperation. The first large, permanent international cooperation in research was CERN.

8. This will not be an entirely new experience. There has been increasing anger, especially among the younger generation of researchers over centrally mandated funding cuts in light of worsening economic crisis. This led, in Hungary, to founding the first independent trade union since Solidarity, after research spending was slashed by 25 percent in December 1987 (*Science,* vol. 240 [27 May 1988], pp. 1142–43). In retrospect, this was an important step in the fall of the regime. Poland's Third Science Congress in May 1986 was also an explicit recognition of the crisis of funding.

9. Cf. the general Gorbachevian/Ryzhkovian reform strategy in the Soviet Union through the summer of 1990.

10. This is true in the countries with a more liberal outward aspect as well as in such repressive states as Romania and Bulgaria. Hungary appears most easily to be effecting leadership changes. In Czechoslovakia, where changes in the academic hierarchy have been more sudden, and in Poland, where the phenomenon of "trimming" is more pronounced, resolution may be further off.

11. This is perhaps most likely in Czechoslovakia, where in distinct contrast to popular rhetoric about democratic traditions, the reality of the past twenty-two years has been one of a repressive regime heavily compromising those with any position or prominence within society.

12. Chemical industry research workers may be especially hard hit as vast tracts of the region's chemical industry are forced to close because of outmoded, inefficient, and severely polluting facilities. If they do remain open, plants will need to convert to more modern, less polluting synthetic processes. For example, East Germany's vast capacity for producing methanol, a low value-added product, may be used to make acetic acid for export rather than to use the old acetylene/acetaldehyde process, which entails use of dangerous mercuric sulfate for catalysis (O'Sullivan and Lepkowski 1990).

13. For many East European scientists this may smack of the old research "problems" approach applied in a hierarchical fashion in Poland in the seventies, under science minister Kaliski, which was viewed as both wasteful of resources and injurious to infrastructure (*New Scientist,* 12 May, pp. 32–33.) Hungary, on the other hand, may once again prove to be in the vanguard in reapproaching

Western practice. In 1987, the new Hungarian Research Foundation instituted a competitive system for awarding grants to fund basic research (*Science*, 15 May 1987, pp. 770–71). Nevertheless, the culture of the past was hard to shake, and there were many charges of the "old boy" system protecting its own.

14. For example, thermodynamics arose as a means for explaining the phenomena accompanying the invention of the steam engine. The vast bulk of research in solid-state physics and in coherent light emission occurred after the inventions of the transistor and the laser. Biotechnology has grown out of molecular biology, to be sure, but that in turn grew out of genetics, which derived from observations based upon practical experience with plant and animal breeding.

15. It would probably be too categorical to associate this view generally with the losers of the Second World War: Japan, Germany, and perhaps even Italy and France. The temptation, however, is there.

16. After the First World War the Soviet Union sought to develop most fully those industrial sectors they identified as being characteristic of the war's winner, the United States, rather than those they identified as being more highly developed in the loser, Germany (see Bailes 1978 for a discussion). Similarly, the apparently science-led mastery of the British and Americans over Germany again in the Second World War may have led to linear conclusions about the role of science in developing technology.

17. One of these was in the field of computer development, generally viewed in the early eighties as a success by most countries with the possible exception of Poland, which felt its own progress was retarded by the cooperative effort.

18. See the interview with G. I. Marchuk, *Pravda*, 29 December 1985.

19. See Popper (1991) for fuller treatment.

20. This will also have an effect on many East European equipment facilities like the laboratory for Low Temperature High Magnetic Field research in Wroclaw, Poland, funded 42 percent from the Soviet Union, and 25 percent and 8 percent respectively from East Germany and Bulgaria.

21. In a single week in May 1990, the Hungarian Academy of Sciences applied to join CERN, the European Laboratory for Particle Physics, the European Space Agency, and hosted a US-USSR-Hungary forum on scientific cooperation and exchange (*New Scientist*, 4 August 1990, pp. 28–29).

22. A study comparing the results of Finnish and Hungarian science shows that in spite of a similar level of research activity and publication, Hungarian science has much less influence as measured by scientometric means than does Finnish. The difference in level of international collaboration is cited as a main cause (Braun, Glaenzel, and Schubert 1985).

23. The MITI (Japan's Ministry for International Trade and Industry) of popular lore, rather than that of historical record, comes to mind as an illustrative example.

24. It should be noted, however, that these changes have been undertaken in a system still adhering in the main to the tenets of central planning, not one where risk taking is likely to receive adequate reward. Even though, in practice, only "twenty out of a hundred projects are found to be successful" (Katsunov, Belyayev, and Bradinov 1983), the planning process assumes a practical return for each. This inclines researchers to choose projects so as to modify risk ensuing from unknown and, by definition, unknowable elements. This leads to less than bold advances and low return on scientific investment.

25. It also suggests a possible source of technological weakness in the standard East European model of separating academy research from the university's education function.

26. This need not be viewed as a one-way street nor as entirely eleemosynary. In return, the Western partner receives the active participation of highly skilled researchers who, because of their very poverty, often excel their Western counterparts in experimental design and sensitivity to the instruments.

27. The United States was of considerable help to both Taiwan and South Korea in building technology infrastructure. Science attachés were active in promoting contacts and provided advice and seed money in founding institutions for research. The demonstration effect from relatively small Western government outlays in Eastern Europe may be similarly profound.

28. A good example is the development of the soft contact lens. The technology was developed in Czechoslovakia. It took a Western commercial partner to recognize the potential and commercialize the result.

29. Various embodied forms of technology will also be transferred. This raises a potential problem for Western government policy as well as an important area for research. However, there is danger for some in the West to overplay the importance of technology transfer in solving Eastern Europe's problems. Among other things this plays into the hands of those who still insist the COCOM embargo was a large source of the region's difficulties and distracts from the true problem, the system itself. Further, the potential for East-to-West technology transfer is frequently ignored.

Bibliography

Bailes, Kendall E. 1978. *Technology and Science under Lenin and Stalin: Origins of the Soviet Technical Intelligentsia, 1917–1941*. Princeton: Princeton University Press.

Braun, T., W. Glaenzel, and A. Schubert. 1985. "Scientometric Indicators of Finnish and Hungarian Publication Performance and Citation Impact in Some Fields of Science (1976–1980)" In K. O. Donner and L. Pal, eds., *Science and Technology Policies in Finland and Hungary*, pp. 255–64. Budapest: Akademiai Kiado.

Cozzens, Susan E., Peter Healey, Arie Rip, and John Ziman. 1990. *The Research System in Transition*. Dordrecht: Kluwer Academic.

Katsunov, S., E. Belyayev, and B. Bradinov. 1983. "Planning for the Advancement of Science in a Socialist Regime." *Science of Science*, no. 3 (11), vol. 3, pp. 225–68.

O'Sullivan, Dermot A., and Wil Lepkowski. 1990. "East Europe Report: Chemical Science," *Chemical and Engineering News*, v. 68 (14 May): pp. 42–61.

Panova, M., and A. Matveev. 1989. *"NTP: Tochki rosta,"* [series] *Ekonomicheskaya gazeta*, nos. 41–44 (October).

Pavitt, Keith and Pari Patel. 1988. "The International Distribution and Determinants of Technological Activities." *Oxford Review of Economic Policy*, vol. 4, no. 4.

Perry, Tekla S. 1990. ". . . So What's to be Done?" *IEEE Spectrum*, vol. 27, no. 10 (October).

Popper, Steven W. 1990. *The Prospects for Modernizing Soviet Industry*. RAND Report R–3785-AF (January).

Popper, Steven W. (1991). *Eastern Europe as a Source of High-Technology Imports for Soviet Economic Modernization*. RAND Report R–3902-USDP.

Teich, Albert. 1990. "U.S. Science Policy in the 1990s: New Institutional Arrangements, Procedures, and Legitimations." In Cozzens et al. (1990), pp. 67–81.

Ziman, John M. 1987. *Science in a "Steady State": The Research System in Transition*. London: Science Policy Support Group.

Eastern Europe and the "Energy Shock" of 1990

John M. Kramer

> In the wake of the first oil crisis, many countries quickly rebuilt their economies along energy saving lines. Eastern Europe didn't hurt because supplies of low cost Soviet raw materials created sheltered conditions and did not motivate the development of energy saving technologies. But today one has to pay for past mistakes. Nothing in the world is free. As Margaret Thatcher says, free cheese is only to be found in a mousetrap.
>
> —Vladimir Voloshin, USSR Academy of Sciences
> *Moscow News*, 1990

> The oil problem is not only a matter of business and dollars, but of stability of a country.
>
> —Vaclav Havel, President of Czechoslovakia
> 7 October 1990.

A new Eastern Europe is emerging from the political upheavals that swept the region in 1989. Naturally, the scope, pace, and significance of change varies among East European states given their diverse cultures, political traditions, and levels of socioeconomic modernization. Yet all regimes in the region—from the crypto-Communist regime in Romania to its democratically elected counterpart in Czechoslovakia—confront many similar challenges that threaten their political and economic stability. How to cope with the "energy shock" of 1990 and beyond wherein the USSR is substantially reducing the volume of, and increasing the price for, its energy exports to the region numbers among the preeminent challenges.[1] The Soviet action "threatens us with collapse," Prime Minister Marian Calfa of Czechoslovakia contends, expressing a widely shared (albeit perhaps hyperbolic) sentiment in Eastern Europe.[2]

Table 1

Eastern Europe: The Energy Gap (thousand barrels per day of oil equivalent)

	Primary energy production[1]				Primary energy consumption[1]				Energy gap[2]			
	1960	1970	1980	1986	1960	1970	1980	1985	1960	1970	1980	1985
Bulgaria	109	151	181	227	131	391	668	715	16.8	61.3	72.9	68.3
Czechoslovakia	761	906	963	987	798	1,136	1,466	1,517	4.6	20.2	34.3	34.9
GDR	1,027	1,172	1,213	1,471	1,166	1,491	1,827	1,978	11.9	21.4	33.6	25.6
Hungary	207	277	278	311	270	431	602	648	23.3	35.7	53.8	52.0
Poland	1,278	1,895	2,446	2,514	1,050	1,653	2,479	2,420				
Romania	487	879	1,102	1,187	376	844	1,338	1,444				17.8
Eastern Europe	3,870	5,280	6,180	6,700	3,790	5,950	8,380	8,720		11.3	26.2	23.2

Source: Compiled from data in Directorate of Intelligence, *Handbook of Economic Statistics* (Washington DC: Government Printing Office, 1987), tables 96 and 97.

[1]Data are for coal, crude oil, natural gas, natural gas liquids, hydroelectric and nuclear electric power expressed in terms of oil equivalent; minor fuels such as peat, shale, and fuelwood are excluded.

[2]Difference between production and consumption as a percent of consumption.

All states of Eastern Europe, except Poland, experience a negative energy gap in which primary energy consumption exceeds primary energy production (Table 1). The energy gap is, by definition, an *ante factum* phenomenon since consumption cannot exceed supply of energy.

East European states have closed their energy gaps since the late sixties primarily through importation from the USSR. The construction of crude oil and natural gas pipelines and a uniform electric power grid in the sixties considerably augmented the capacity of the Soviet Union to export to Eastern Europe. Eastern Europe has imported almost all of its natural gas, 80 percent of its imports of crude oil and petroleum products, and more than 70 percent of its imports of hard coal from the Soviet Union. In 1985, all East European states except Romania derived more than 75 percent of their total imports of energy from the Soviet Union. The respective figure for Romania—reflecting its pursuit of independence from Moscow during the regime of Nicolae Ceausescu—was 19 percent.

These imports comprised an integral component of the Stalinist model of economic development that the Soviet Union imposed on Eastern Europe after World War II. In brief, this model sought economic development primarily through *extensive* growth (quantitative increases in factor inputs) rather than *intensive* growth (qualitative increases in factor productivity). Its well-known priorities included the forced draft development of energy-gobbling heavy industries, including steel, chemicals, and mining. The goal was to maximize, not minimize, the utilization of energy as this resource became a principal sinew of the new industrial infrastructure. That all East European states consume on average between 30 and 50 percent more energy than do industrialized capitalist countries to produce similar units of national income largely derives from this circumstance.[3]

Soviet energy provided essential underpinning to the economically inefficient East European economies and their pursuit of extensive growth. East European economies became addicted to this energy, which they acquired primarily for "soft" goods, i.e., goods uncompetitive on international markets. Eastern Europe concomitantly became isolated from the competitive international market and integrated into the technologically backward socialist market. These conditions were tolerable as long as Eastern Europe was assured of steadily increasing volumes of Soviet energy. The energy shock of 1990 represents the end of the era of cheap and plentiful energy from the USSR. As a

Soviet source colorfully phrases it, the Soviet Union is forcing Eastern Europe to go "cold turkey" by ending the "oil fix" to it.[4]

What this circumstance portends for the future welfare of the East European polities constitutes the focus of this study. We explore this issue by examining first the origins and then the present and likely future status of the Soviet-East European energy relationship.

The Energy Relationship: 1960–85

Tension, conflict, and the vigorous pursuit of national interests characterized the Soviet-East European energy relationship from its inception.[5] Debate revolved primarily around the interrelated issues of the amount of and prices for Soviet energy to members of the "socialist commonwealth."

In particular, observers both within and without the USSR repeatedly expressed skepticism that indigenous energy production was sufficient for the Soviet Union to satisfy internal demand, to meet requirements for convertible ("hard") currency through energy exports to the world market, and to continue as the primary source to close East Europe's energy gap. By the late sixties, Soviet commentators already argued that escalating costs incurred in exploiting energy resources located increasingly in remote and inhospitable regions of western Siberia made their export to Eastern Europe too expensive.[6] However, an Eastern Europe less dependent on the Soviet Union for vital raw materials—and commensurately less susceptible to political pressure—was probably unappealing to Soviet leaders after the turmoil in Czechoslovakia in 1968.

The opportunity costs to the USSR of its energy exports to capitalist and socialist states also influenced its policy. Exports of energy to the former states are the Soviet Union's primary source of hard currency. In contrast, considerable controversy exists over whether the USSR or Eastern Europe benefited more from their trade in energy.

Between 1958 and 1975, prices in Comecon trade followed the so-called Bucharest formula—that is, prices for individual commodities were established for a five-year period and based on the average world-market price for that commodity in the preceding five-year period. The rapid escalation in world-energy prices after the October 1973 war between Israel and the Arab states permanently ended the Bucharest formula. The USSR demanded an immediate revision of the

Table 2

Eastern Europe: Estimated Price of Soviet Oil
(dollars per barrel)

Year	World market price	Comecon price	Comecon price as percent of world price
1978	12.70	11.39	90
1979	17.26	13.89	80
1980	30.22	14.01	46
1981	32.50	15.67	48
1982	34.00	21.91	64
1983	29.00	25.89	89
1984	28.20	27.28	97

Source: Radio Free Europe, *Background Report*, no. 155 (24 August 1984).

formula, since by March 1974 the Comecon price for oil was already 80 percent below the comparable world-market price. The ensuing revision entails annual adjustments in prices based on average world-market prices for individual commodities in the preceding five-year period. These changes directly affected the official Comecon price for oil between 1978 and 1984 (Table 2).

If one interprets these data literally, prevailing prices for oil in Comecon trade represented an economic windfall for Eastern Europe. Eastern Europe actually benefited even more than these data indicate. In reality, Table 2 overstates the price of Soviet oil because it uses the artificially high official rate to convert rubles into dollars. Hungarian economists utilizing more realistic exchange rates estimated in 1984 that the actual price of Soviet oil to Eastern Europe was approximately 60 percent below the prevailing world-market price.[7]

The USSR also contended (accurately) that Eastern Europe paid for this energy mostly with soft goods, whereas the USSR received high-quality ("hard") goods or convertible currency from capitalist states for this commodity. Eastern Europe continued this practice despite repeated demands from the USSR to end it.[8]

However, Eastern Europe contended (also accurately) that several factors substantially increased the price of Soviet energy. First, Eastern Europe claimed that the Soviet Union forced it to buy soft goods at inflated prices to compensate the latter for losses it allegedly incurred in energy sales. Second, the Eastern Europeans purchase above-plan deliveries of—and Romania purchases all of its—petroleum from the Soviet Union at world-market prices.

Third, the Soviet Union imposed a de facto increase in price on its energy exports by tying their availability to East European participation in the development of energy resources located in the Soviet Union.[9] Participation in these so-called joint projects represented a hardening in terms of trade because it obligated Eastern Europe to supply the Soviet Union with hard goods either produced domestically or purchased on capitalist markets with repayment in the product of the project. The creditor also extends to the debtor in these projects (typically Eastern European states and the USSR, respectively) an implicit subsidy through artificially low rates of interest (2 to 4 percent) charged on credits. Then, too, official prices of the debtor state are used to value credits in joint projects. Eastern Europeans complain bitterly that the ensuing valuations are arbitrary and substantially understate the monetary contribution they make to joint projects.[10] Creditors also express concern over provisions for repayment in joint projects. The policy of valuing repayment at the time of delivery prevents creditors at the beginning of a project from determining actual rate of return on investment. After repayment of credits, creditors retain no right (as is usual in international practice) to preferential terms of purchase from completed projects.

These disputes only began to affect the supply of Soviet energy in the eighties. Soviet energy exports to Eastern Europe increased at annual rates of 9.5 percent and 6 percent, respectively, in the periods 1971–75 and 1976–80. The information publicly available indicated that these exports would continue to increase during the period 1981–85. For example, Erich Honecker, the East German Communist party leader, speaking at the Party's national congress in April 1981, asserted that annual deliveries of oil from the Soviet Union "have been securely agreed upon on a long-term basis." Similarly, the secretary general of Comecon reiterated that "under no circumstances" would Moscow reduce oil deliveries to Eastern Europe before 1985.[11]

Then, suddenly, in 1982 the Soviet Union announced an approximately 10 percent cut in its planned oil deliveries to Eastern Europe between 1982 and 1985. The dramatic drop in world-market prices for energy after 1981 largely explains the Soviet action. As one Western analyst observes, the Soviet Union responded to declining prices by "frantically trying to remain in the market, even at the expense of its allies."[12] Thus, in 1982 and 1983, it increased its exports of crude oil and petroleum products to capitalist states by 25.9 and 17.4 percent, respectively, while simultaneously enacting the aforementioned reduction in deliveries to Eastern Europe. This action provided a harbinger of the even greater problems that awaited Eastern Europe after Mikhail Gorbachev assumed power in the USSR in March 1985.

The "Energy Shock" of 1990

In the USSR itself, Gorbachev's drive to revitalize the ailing economy, major problems besetting the energy sector, and declining revenues from energy exports to the international market have combined with the well-known political events in Eastern Europe since 1989 to make the future of Soviet energy exports to the region highly problematic.

Gorbachev initially sought to create a self-styled "unified socialist market" among the socialist states based on the mutual development and exchange of world-class commodities and technologies.[13] Altering what Soviet premier Nikolai Ryzhkov labelled the "archaic" commodity structure of Soviet trade with Eastern Europe constituted a key component of the Gorbachev program.[14] The proposed alteration entailed substantially increasing the proportion of machines and other industrial manufactures in Soviet exports while simultaneously receiving more hard goods from Eastern Europe in payment for energy. If these objectives remained unrealized, a Soviet commentator prophetically predicted, "then the very foundation of economic cooperation between socialist states will be severely damaged, and this is fraught with the most serious dangers that are not only economic in nature."[15]

In fact, the Soviet desiderata remained unfulfilled. Consequently, during 1987–89 the value of total trade turnover between the Soviet Union and Eastern Europe stagnated.[16] The value of Soviet exports to the region in this period declined substantially, primarily because of declining prices for energy and raw materials in Comecon trade. One Soviet economist calculated that in 1988 the USSR would have earned an additional $4 billion if it had sold on the international market the volume of oil it delivered to Eastern Europe. A Western source estimates that in 1989 Eastern Europe paid only 39 percent of the world-market price for a barrel of Soviet oil.[17] Such losses are "criminal," one official charged, given the depressed state of the Soviet economy.[18] In these circumstances, the Eastern Europeans presumably

were gratified that the USSR (reportedly after acrimonious negotiations) agreed to maintain its current level of oil exports to them
through 1995.[19] There were even indications that the USSR might
export more natural gas than agreed upon between 1991 and 1995.[20]

These commitments are now void. First, the USSR will not meet its
planned target for exportation of crude oil to Eastern Europe in
1990. It substantially underfulfilled this target in the first half of 1990
to Bulgaria, Czechoslovakia, Hungary, and Poland. Then, Premier
Ryzhkov announced at the 28th Congress of the CPSU in June that
shortages of fuel on the domestic market (especially in agriculture)
compelled the USSR to reduce its exports of crude oil, including those
to Eastern Europe, by 7 MTs (million metric tons). Subsequently, a
Soviet source reported that "all the signs" now indicate that in 1990 the
USSR will export to Eastern Europe approximately 20 percent less
crude than it did in 1989.[21] This report belies repeated Soviet assuran-
ces that it would increase exports of crude in the latter part of 1990 to
compensate for the earlier shortfall in deliveries.[22]

Preliminary data indicate that in 1990 compared to 1989 the actual
decline in Soviet oil deliveries to Eastern Europe may be closer to 25
percent than the aforementioned estimate of 20 percent. An estimate
for 1991, which excludes Romania where data are lacking, suggests
that these deliveries could decline by another 22 percent compared to
1990. (Table 3)

The tentative nature of these estimates must be stressed. First, the
Eastern Europeans—presumably acting upon information received
from the USSR—have repeatedly revised their estimates of the volume
of crude oil they will receive from the Soviet Union in 1990. For
example, Deputy Premier Vladimir Dlouhy of Czechoslovakia estimated in July that his country would receive 14.07 MTs of crude from
the USSR in 1990. In October, Czechoslovakia's deputy minister of
foreign trade provided the estimate of Soviet oil deliveries for 1990
included in Table 3. Shortly thereafter, the press secretary for Czechoslovak president Vaclav Havel stated that Czechoslovakia might receive barely 9 MTs of crude from the USSR in 1990.[23] Poland
provides another example. Jerzy Osiatynski, minister for the Office of
Central Planning in Poland, contends that the USSR officially informed Poland in August 1990 that in 1991 deliveries of crude oil
would "amount to zero." The vice minister for Foreign Economic Cooperation in Poland subsequently indicated that the Soviet Union

Table 3

Eastern Europe: Estimated Volume of Crude Oil to be Imported from the USSR 1990, 1991 (millions of metric tons)

	1989 (actual)	1990 (esti-mated)	Percent change (1989–1990)	1991 (esti-mated)	Percent change (1989–1991)
Bulgaria	11.46	9.69	−15.4	7.20	−37.2
Czechoslovakia	16.60	12.60	−24.1	8.00	−51.8
Hungary	6.32	5.09	−19.5	4.00	−36.7
Poland	13.03	9.12	−30.0	8.00	−38.6
Romania	3.94	3.40	−13.7	NA	NA
Eastern Europe	52.35	39.90	−23.8	28.20	−46.1*

*Estimate excludes Romania.

Sources: Data for 1989 from Plan Econ Inc., *Plan Econ Report*, 8 June 1990. Data for 1990 and 1991:

Bulgaria—from *Duma*, 29 July 1990, in Foreign Broadcast Information Service, *East Europe Daily Report* (FBIS-*EEDR*), 2 August 1990, p. 10; Sofia Domestic Service, 28 September 1990, in FBIS-*EEDR*, 1 October 1990, p. 18;

Czechoslovakia—from *Svobodne slovo*, 2 October 1990, in FBIS-*EEDR*, 9 October 1990, p. 20; *Hospodarske noviny*, 28 September 1990, in FBIS-*EEDR*, 4 October 1990, p. 13;

Hungary—from Magyar Taviruti Iruda (MTI), 3 September 1990, in FBIS-*EEDR*, 4 September 1990, p. 25; MTI, 11 September 1990, in FBIS-*EEDR*, 13 September 1990, p. 29;

Poland—from *Der Standard* (Vienna), 13 September 1990, in FBIS-*EEDR*, 14 September 1990, p. 1; Polska Agencya Prasowa, 17 October 1990, in FBIS-*EEDR*, 18 October 1990, p. 28;

Romania—from *Pravda* (Moscow), 22 October 1990.

would deliver 8 MTs of crude oil to Poland in 1991.[24] A Hungarian official informs that in "preliminary talks" the USSR has offered to supply Hungary with 4 MTs of crude in 1991. Significantly, however, Hungary has devised an emergency plan to ration energy supplies if its oil imports from the USSR are "terminated."[25]

The devolution of decision-making authority in the USSR from Moscow to the union republics engenders additional uncertainty. As Prime Minister Calfa of Czechoslovakia expressed it, "the republics are bickering with the center and no one knows what will happen" with supplies of oil for export.[26] Czechoslovakia recently concluded a bilateral contract with an oil-production firm in Tyumen Oblast (Russian Republic) to purchase 500,000 tons of crude oil in 1990. Czechoslovakia will pay for the oil with goods, including trucks, buses, bulldozers, and a "large batch" of consumer goods. Czechoslovakia hopes to conclude similar agreements in the future for upward of 3 MTs of crude annually. The agreements could entail direct Czechoslovak participation in the exploitation of energy reserves located in the Russian republic.[27]

Second, uncertainty attends the future status of exports of natural gas and electricity from the USSR. Most Soviet natural gas flows to Eastern Europe as repayment for credits the latter extended to exploit gas deposits in the Soviet Union. Czechoslovakia, Hungary, and Poland had hoped—in part, to mitigate their intensive environmental pollution by utilizing natural gas at the expense of highly polluting soft coal—to increase substantially importation of this fuel even before the USSR completed repayment of credits in the late nineties.[28] These may prove abortive hopes. Poland recently announced that in 1991 it would receive approximately 23 percent less natural gas from the USSR than in 1990.[29] There also do not exist agreed-upon terms of trade to purchase natural gas from the USSR after it completes repayment of outstanding credits. Hungary, which held the "impression" that it would double Soviet natural gas imports by the year 2000, has begun "urgent negotiations" with the USSR to clarify future terms of trade.[30] Finally, a Soviet source claims that electricity exports to Eastern Europe have "no future," in part because cutbacks in the Soviet nuclear program since the disaster at the Chernobyl' nuclear plant have reduced indigenous capacities to meet demand for power.[31] Hungary is the East European state most dependent upon these exports. Hungarian officials acknowledge the "specter hovering over our heads" of reductions in, or even termination of, electricity imports from the USSR.[32]

Third, the USSR now demands that its trade with Eastern Europe, from January 1, 1991, be conducted in convertible currency using world-market prices.[33] Convertible currency would replace the transferable ruble, an accounting device transferable in name only used within Comecon to value goods exchanged in barter trade. One Soviet commentary, hailing this initiative, characterized the transferable ruble and attendant system of barter trade as "ponderous and destructive" and "doomed to fail" because of its "violence to economic laws." Similar sentiments have long been expressed in official and academic circles in Eastern Europe.[34]

Yet the looming economic consequences of the Soviet policy may compel changes in its substantive provisions and timetable of im-

Table 5.4

Eastern Europe: Estimated Cost of Crude Oil Imported from USSR
(1990, 1991)

	1990 (in billions of dollars)	1991 (in billions of dollars)
Bulgaria	2.180	1.620
Czechoslovakia	2.835	1.800
Hungary	1.145	0.900
Poland	2.053	1.800
Romania	0.765	NA

Sources: Figures are based on estimated volumes of Soviet crude oil deliveries contained in Table 3 and on an estimated world market price of $225 per ton ($30 per barrel). See note 36 for a discussion of the rationale for, and the caveats associated with, using $30 per barrel as the prevailing world market price for oil.

plementation. As a close adviser to Poland's Premier Tadeusz Mazowiecki warned, an immediate switch to trade in convertible currency could "lead to a break in economic ties" between the USSR and Poland.[35] The remark applies equally to other states in Eastern Europe (Table 4).

What the Eastern Europeans expected to pay in dollars at world-market prices for the estimated volume of crude the USSR exported to them in 1990 and 1991 can be judged from the figures in Table 4.[36] In calculating the total economic impact of the new pricing policy, one must also consider that it applies to all Soviet exports to Eastern Europe (which mostly comprise raw materials that are "hard goods" in international trade) and not just oil. In contrast, East European exports of manufactures to the USSR would suffer disproportionately under the new policy because they consist primarily of "soft goods" that either cannot be sold or can be sold only at subsidized prices on world markets. A Soviet source estimates—although Western analysts dispute this—that Comecon prices for machinery and equipment are roughly 50 percent higher than comparable world-market prices.[37]

The Eastern Europeans have sought to mitigate the adverse consequences of the new pricing policy. First, they have appealed to Western governments for financial aid to this end. Conflicting reports exist on the likelihood of such aid. One Western source reports that officials of the European Economic Community "responded favorably" to the request of Prime Minister Jozsef Antall of Hungary to establish a special reserve fund of credits that East European states can draw upon to ease the transition to hard-currency trading with the USSR. But another source informs that a similar proposal made by Czechoslovakia "has neither been accepted nor is being further considered by the European Economic Community or any other group at present."[38]

Second, the Eastern Europeans have lobbied vigorously for a "phased" introduction of the new pricing policy with barter trade in raw materials continuing in the interval. Justifying such proposals with a widely held view in Eastern Europe, a Hungarian source argues that the legacy of Soviet rule in the region makes it "essential" that the USSR "share" the costs of overcoming it. Their economies served "Soviet requirements," this source contends, leading to an "outdated production structure" whose manufactures "have fallen way behind world standards." This argument reportedly has as yet made "no progress" with Soviet officials.[39]

Third, Czechoslovakia, Hungary, and Poland have pressed the USSR to convert its ruble debt to them into dollars that could then be used, *inter alia,* to purchase Soviet energy. The rate of exchange to be used in this conversion has prevented an agreement on this issue with Czechoslovakia and Poland. Reportedly, after talks that were "indeed tough," the USSR agreed to convert Hungary's ruble surplus for 1989 into a credit of $720 million. However, a Czechoslovak source contends that the two sides have not concluded such an agreement. The East European states are now considering a "joint" demarche toward the USSR to resolve the conflict.[40]

Yet a compromise over pricing appears likely—at least with some countries. Konstantin Katushev, USSR minister of foreign economic relations, informs that barter trade through "mutually balanced deliveries" will continue in Comecon for a "certain category of commodities."[41] The Czechoslovak-Soviet trade pact for 1991 reflects this circumstance. The pact provides, *inter alia,* that Czechoslovakia will pay "with goods" for Soviet "raw materials deliveries." Commenting upon the pact, Deputy Premier Vaclav Vales of Czechoslovakia reported without elaborating that this "transitional state" of barter trade "may continue in the future."[42] The USSR and Hungary have reached a similar agreement on barter trade.[43]

Both Soviet and East European commentators question what is "new" about a policy that continues barter trade for the overwhelming share of Soviet exports to Comecon countries.[44] The nascent status of the proposed pricing policy permits only tentative answers to this question. First, prevailing world-market prices (probably expressed in dollars) will be used to value bartered goods, thereby eliminating the one-year lag of Comecon prices behind world prices. As noted, this will result (all things being equal) in a substantial deterioration in East Europe's terms of trade with the USSR as the price of the latter's exports to the region increase substantially. For example, Soviet calculations—which Hungarian analysts do not dispute—indicate that utilizing world-market prices in 1991 will cost Hungary approximately $1.5 billion if the structure of Soviet-Hungarian trade remains unchanged. An estimate in July 1990 (when world-market prices for oil were around eighteen dollars per barrel) calculated that world-market prices would make Soviet oil 3.5–4 times more expensive for Hungary.[45] A ranking Czechoslovak official in the Ministry of Foreign Trade estimates that Czechoslovakia must export 20 percent more goods in 1991 to receive the same quantity of "selected" raw materials from the USSR as in 1990.[46] It is unclear precisely what "selected" raw materials his estimate applies to or whether the estimate is based upon the planned or *probable* volume of Soviet deliveries in 1990.

Second, the USSR is increasing the volume of its energy exports that it will sell to Eastern Europe only for hard currency or hard goods.[47] As noted, Eastern Europe traditionally has purchased above-planned deliveries of oil from the USSR with convertible currency at world-market prices, although comprehensive data are lacking on the magnitude of this trade.

Third, the new policy permits participants to reject low-quality goods offered in barter trade. Since this provision would apply primarily to East European manufactures rather than Soviet raw materials, it could serve to increase the overall quality of East European exports to the USSR.[48] Doing so would reverse the long-standing practice of the Eastern Europeans paying only lip service to Soviet demands for higher-quality goods in barter trade. The USSR may be far less tolerant of such lip service emanating from non-Communist regimes than from their Communist predecessors.

Soviet Motivations

Despite the manifold ambiguities, uncertainties, and contradictions in the evolving Soviet energy-export policy toward Eastern Europe, one fact seems clear: Eastern Europe will soon be paying more (whether in convertible currency or commodities) to receive a smaller volume of energy from the USSR. A suggestive list of interrelated factors illuminates the rationale for this policy.

First, shortages of fuel on the domestic market—the professed reason for reducing oil exports—at best explains the timing of the announced cutbacks. "Unjustifiably high" energy exports to Eastern Europe are one reason the economy experiences "energy hunger and equipment stands idle," a prominent academic critic of these exports contends.[49] Shortages of gasoline and diesel fuel in agriculture especially hampered the harvest of 1990.[50]

Yet these shortages (no matter how disruptive in the short run) are only symptomatic of deep-seated problems besetting the energy sector that limit the volume of energy available for export. Prospects for petroleum production are especially worrisome. Production decreased by 3 percent in 1989 (to 607 MTs) versus 1988 and dropped another 5 percent in the first half of 1990. The USSR likely will extract around 585 MTs of crude in 1990 (in contrast to the original plan target of 635 MTs), although several non-Soviet sources believe extraction could even fall below 500 MTs.[51] Overall, the USSR produced 2 percent less energy in the first half of 1990 than in the comparable period of 1989.[52] Analyzing the reasons for the decline in oil production lies beyond our purview. Suffice it to say that these include poor management, limited discovery of new reserves, outdated exploration and production technology, peaking and declining output at major fields, and severe labor shortages and unrest among workers in the key oil-producing regions of western Siberia.[53] Preliminary agreements recently concluded with Western oil companies (including Texaco and Chevron) for joint participation in exploration and exploitation of energy reserves could, in the long run, mitigate some of these problems through an infusion of foreign expertise and technology.[54]

Second, these difficulties exist against the backdrop of declining hard-currency earnings from exports of energy. As noted, the USSR derives the overwhelming share of its hard currency from these exports. In both 1986 and 1987, the USSR established records for volume of energy exports to nonsocialist states, but earned only $13.5 billion and $16 billion, respectively, from these exports (compared to $23.6 billion in 1983) because of declining world-market prices for energy.[55] In 1989—compared to 1988—the USSR exported approximately 20 MTs less crude oil and petroleum products and earned approximately 1.5 billion rubles less from these exports.[56] The ensuing shortage of hard currency has tarnished the USSR's previous reputation for scrupulously fulfilling its financial obligations to Western creditors. The USSR has even earmarked one-half of a five-billion DM credit from West Germany to pay bills long overdue to German firms.[57]

In these circumstances, continuing the "petrodollar rain" whose "mighty torrents" have "irrigated our allies's economic ground for almost two decades" held little economic attraction.[58] The new pricing policy with its emphasis on profitability and market-determined prices is also consistent with the imperatives of Mikhail Gorbachev's perestroika of the Soviet economy.[59]

Third, the demise of Communist regimes undermined the political rationale for propping up East European economies with huge amounts of energy sold at subsidized prices. This rationale, a Soviet source candidly admits, entailed using energy exports to help Eastern Europe "build socialism Stalinist style."[60] It may be correct, as Soviet and East European spokesmen repeatedly assert, that the export cutbacks are not in retaliation for the independent policies of the region's non-Communist regimes.[61] Yet these regimes cannot credibly argue, as did their Communist predecessors, that the USSR has vital ideological and political interests in their survival that justify incurring substantial economic losses in energy trade.[62] Sound commercial relations in the best traditions of capitalism, not political retaliation, must become the leitmotiv of the new energy policy toward Eastern Europe, Soviet officials probably reason.

This reasoning especially applies to the lucrative hard-currency trade Eastern Europe conducted by reexporting imported Soviet crude oil in refined form. Facilitating such trade makes little economic or political sense when countries such as Hungary (which sold approximately $400 million worth of petroleum products on West European markets in 1989) announce they are leaving the Warsaw Pact and could conceivably even enter NATO.[63]

Fourth, the process of devolving political power to the union republics may be limiting the capacity of central authorities in the USSR to make binding long-term commitments for the export of energy resources. All declarations of sovereignty issued to date by union republics assert the right to control fully the disposition of natural resources located on their territories. Will other East European states emulate the example of Czechoslovakia and negotiate directly with union republics and their constituent political subdivisions to purchase energy resources? How successful Eastern Europe will be in this endeavor represents a crucial variable in the future Soviet-East European energy relationship.

Finally, the USSR will permit some barter in its energy trade with Eastern Europe. This circumstance suggests that the USSR retains important political and economic interests in the region, which militate against a complete break in trade with its former satellites. Politically, it would be absurd to argue that the demise of East European communism makes the USSR indifferent to political developments in the region. For example, the USSR would hardly welcome an anti-Soviet regime in Poland stirring up national sentiments among the ethnic Polish population of Soviet Ukraine, or its Romanian counterpart pursuing similar activities among the ethnic Romanian population of Soviet Moldavia. States that remain dependent upon the USSR for vital supplies of energy—even if they pay higher prices and provide better-quality goods for reduced volumes of these supplies—are unlikely to pursue such actions. President Havel has explicitly raised the possibility of political retaliation if the USSR reneges on its energy commitments to Czechoslovakia. In this situation, Havel asserts, Czechoslovakia would not display "excessive consideration" for Soviet interests regarding the future of the Warsaw Pact and the emerging security order in Europe.[64]

Economically, the USSR has made clear its interest in trade with Eastern Europe if it can increase the importation of foodstuffs and better-quality consumer goods and decrease the importation of outdated machinery.[65] Realistically, a Soviet source contends, the USSR and Eastern Europe are "doomed" in the near term to mutual trade, because both sides lack requisite quantities of high-quality goods and convertible currency to trade extensively on international markets.[66] This circumstance suggests that the USSR may seek primarily to pressure Eastern Europe into supplying it with higher-quality goods in trade when it threatens a possible termination of oil exports to Poland and similar draconian measures. Eastern Europe also retains a vested economic interest in importing Soviet energy even at world-market prices in hard currency. For example, importing energy from the nearby USSR through a system of already operating oil and natural gas pipelines entails substantially lower costs in transportation and infrastructural facilities for Eastern Europe than similar purchases from distant suppliers in the Middle East and Latin America.[67]

And yet the USSR *does* export one commodity to Eastern Europe—oil—which it also sells readily on the international market. Could the USSR export to this market a significant proportion of the oil it now sells to Eastern Europe? The state of the international oil market largely determines the answer. Its current state, with spiraling prices precipitated by Iraq's invasion of Kuwait, makes this far more economically feasible than it would have been in the late eighties when world-market prices for oil were so depressed. The USSR could easily earn an additional $1.5 billion from oil exports through the remainder of 1990 solely through higher world-market prices for oil.[68] The temptation to earn even more hard currency by diverting oil from Eastern Europe to the world market places additional pressure on Eastern Europe to satisfy the demands of the USSR for higher-quality goods in mutual trade. The potential for diversion refutes analysts who contend there is "no way" the USSR can increase oil exports to the world market.[69]

Prospects

The volatile state of the Soviet–East European energy relationship permits only tentative judgments about its impact on future political and economic life in the new Eastern Europe. The available evidence suggests its impact may be more deleterious in the short run than in the long run.

In the short run, cutbacks in Soviet energy deliveries are disrupting production in key industries (e.g., petrochemicals), exacerbating economic decline in the region and generating political turmoil as angry consumers vehemently protest higher prices for fuels and electricity designed to pay for, and conserve, more expensive imported energy.[70] Eastern Europe clearly will soon devote a greater share of its national wealth to pay for importation of energy, which (all things being equal) will constrain its capacity to pursue other investment opportunities, including those intended to promote consumer welfare and the technological modernization of its obsolete industrial infrastructure. A Western source prophesies that the East European economies risk "collapsing into destitution" from the energy shock of 1990. This collapse would have profound political implications: "Governments will be hard pressed to maintain order; national rivalry between the Czechs and Slovaks, between the Romanians and Transylvanian Magyars, and between the Bulgarians and Bulgar Turks, for instance, would probably explode."[71]

Three variables will help determine whether this is a hyperbolic or realistic scenario. First, how demanding will the USSR be in linking its energy supplies to higher-quality exports from Eastern Europe? Eastern Europe—in the short run—possesses limited capacities to increase the production of hard goods. Consequently, a "Catch 22" situation could emerge wherein to satisfy Soviet demands Eastern Europe diverts goods now exported to capitalist markets—and thereby limits its capacities to acquire the sophisticated technologies its economies so desperately need and, concomitantly, to produce the very hard goods the USSR so desires. Further, the USSR especially wants Eastern Europe to supply it with precisely those commodities—high-quality consumer goods and foodstuffs—most desired by East European populations, themselves increasingly impatient for the "good life" of consumer plenty. That Czechoslovakia has now established a crisis-management team to cope with the emergency situation in Soviet oil supplies testifies to the gravity of the situation. President Havel will soon hold a personal "oil summit" with President Gorbachev and Prime Minister Ryzhkov to plead for an increase in oil deliveries from the USSR.[72]

Another relevant variable involves the volume of, and terms of trade for, oil that Eastern Europe will import from the international market to compensate for less Soviet crude. In this context, Iraq's invasion of Kuwait, the ensuing United Nations imposed trade embargo with Iraq, and the spiraling world-market prices for oil engendered by the Gulf crisis represent serious blows for Eastern Europe. As Vaclav Klaus, Czechoslovakia's minister of finance, told officials from the World Bank and the International Monetary Fund assembled to discuss the economic hardships created by the Gulf crisis, "My English does not offer me adjectives for describing the problem now."[73]

Iraq appeared the most likely source before the embargo to compensate Eastern Europe for the shortfall in Soviet deliveries. This status derived from Iraq's willingness to pay off its considerable debt to East European states mostly with oil.[74] There ensued "oil for debt" deals with Bulgaria (4.75 MTs between 1990 and 1994), Poland (2–3 MTs in 1990–91),

Hungary (210,000 tons in 1990), and Czechoslovakia (for an amount "to cover the deficit in deliveries from the Soviet Union").[75] Eastern Europe was also involved in a complex three-way "oil for debt" deal wherein Iraq would export 10.7 MTs of crude in 1990 to the USSR, which would then reexport this oil to Eastern Europe.[76]

The embargo against Iraq has compelled Eastern Europe to seek alternative sources to supply oil. Italy and Venezuela have indicated their willingness to play this role, but will sell crude only for hard currency.[77] For a country such as Bulgaria—which has suspended both principal and interest payments on its hard-currency debt—this is a prohibitive condition. Hungary, with one of the highest per capita foreign debts in the world, will purchase upward of 650,000 tons of crude on the international market (mostly from Algeria and Libya) in 1990 for approximately $200 million.[78] Eastern Europe's chronic shortage of hard currency makes it imperative that it obtain much of its oil from the international market in barter trade. Iran appears particularly willing to conclude such deals. Iran recently agreed to supply Bulgaria with 1 MT of crude oil in 1991 in exchange for Bulgarian machine tools, trucks, cigarettes, electrical equipment, and refrigerator compressors. Reportedly, Iran is willing to supply Czechoslovakia on a long-term basis with between 5 and 10 MTs of crude annually with payment in goods. Iran will accept payment in barter for 400,000 tons of oil it will export to Czechoslovakia in 1990. Similarly, Poland will purchase 500,000 tons of crude from Iran in 1991 and perhaps as much as 2.5 to 3 MTs in 1991, although terms of trade have not been announced.[79] If such deals actually materialize, they will significantly mitigate the effects of the energy shocks Eastern Europe has felt in 1990 from the USSR and the crisis in the Persian Gulf.

The magnitude of assistance to this same end that Eastern Europe will receive from Western states constitutes a final relevant variable. As noted, East European states have made several so far unsuccessful demarches to the European Economic Community seeking such aid. The International Monetary Fund and the World Bank have promised an, as yet, unspecified amount of assistance to help compensate these states for losses they have incurred in supporting the embargo.[80] The USSR could also extend such assistance, the foreign minister of Hungary recently suggested, by selling increased volumes of oil to Eastern Europe on favorable terms.[81] To date, the USSR has not responded publicly to this suggestion.

The long-run prospects are potentially brighter. Most fundamentally, the end of the era of cheap Soviet energy may compel Eastern Europe to enact decisive measures to conserve energy with concomitant benefits throughout their economies.[82] As noted, all East European states consume between 30 and 50 percent more energy than do industrialized capitalist states to produce similar units of national income. A Soviet source identifies how the USSR contributed to this circumstance: "For many years the East European countries' energy-supply problem was solved mainly by deliveries of cheap Soviet fuel. In these conditions, the policy of energy conservation was only proclaimed, but . . . never carried out. Today these countries are having to pay for this."[83]

An effective program of energy conservation must include wholesale and retail prices for energy resources that reflect marginal cost. For several reasons—the Marxist law of value that considers natural resources "free goods," a desire to encourage the production of energy-intensive commodities, and a concern to avoid political unrest among the population—Communist regimes traditionally attached prices to energy resources that were too low to stimulate conservation. Indeed, the Stalinist economies of Eastern Europe, with their pursuit of extensive growth, were largely devoid of any economic stimuli to use energy and other resources efficiently. East European regimes already have responded to the energy shock of 1990 by enacting substantial increases in prices for energy resources.[84]

Energy prices reflecting marginal cost will produce other manifest benefits as they accomplish their primary goal of reducing consumption of energy. They will promote technological modernization by making it economically more rational to utilize advanced energy-saving technologies when producing and consuming energy. An obvious example of this circumstance would be if consumers coping with soaring prices for gasoline demanded more fuel-efficient automobiles. They will free desperately needed monies for other investment objectives by reducing the disproportionate share of investment resources in industry (between 40 and 45 percent throughout Eastern Europe) now devoted to produce energy.[85] They will mitigate the intensive pollution of the environment in Eastern Europe—perhaps the preeminent social pathology in the region—by promoting technological modernization and providing an economic incentive to conserve resources now emitted as pollutants.[86] Finally, they will spur the overall marketization of the economies being pursued with varying degrees of enthusiasm by regimes in the region. Marketization constitutes a necessary, albeit not sufficient, condition to enhance economic efficiency and consumer welfare in Eastern Europe.

Enhanced economic efficiency and technological modernization will become imperative as the crisis in Soviet-East European energy relations undermines the latter's links with the technologically backward Soviet market and concomitantly spurs its links with the technologically advanced Western market. Eastern Europe either will become technologically competitive on the international market or find itself increasingly unable to import from this source requisite quantities of energy resources and other vital commodities. Coping successfully with the challenge of international competitiveness will intensify economic and political interaction between Eastern Europe and Western states. Only these states possess the fiscal resources and technological capabilities to assist Eastern Europe in shedding the legacy of technological obsolescence inherited from the USSR. Czechoslovakia, Hungary, and Poland appear to be the East European states best able to exploit Western assistance to this end.

Ironically, then, the end of the era of cheap Soviet energy—which has generated such understandable anguish and concern in the region—may in the long run impel Eastern Europe toward greater technological modernization, international competitiveness, and economic and political rapprochement with the West. These manifest benefits will not come without considerable economic and political costs. In the short run, inflation, unemployment, social tension, and political polarization are seemingly inevitable concomitants of the transition from Stalinism to political pluralism and marketization in Eastern Europe. The new Soviet energy policy toward Eastern Europe may offer the fledgling post-Stalinist regimes in the region no choice but to endure these costs. Eastern Europe may eventually thank the USSR for forcing it to go "cold turkey" and end its addiction to Soviet energy.

Notes

1. For a detailed analysis of the energy policies of the East European regimes, see John M. Kramer, *The Energy Gap in Eastern Europe* (Lexington, MA: Lexington Books, 1990). Eastern Europe herein refers to Bulgaria, Czechoslovakia, Hungary, Poland, and Romania. East Germany is not considered because of its reunification with West Germany.

2. *Mlada fronta dnes* (Prague), 2 October 1990, in Foreign Broadcast Infor-

mation Service, *East Europe Daily Report* (hereafter FBIS-*EEDR*), 10 October 1990, p. 21.

3. For example, a Soviet source reported in 1987 that the European members of Comecon "consume approximately 40 percent more power per unit of output than do European Economic Community countries" (*New Times* [Moscow], 12 January 1987, p. 32). *Planovoe hospodarstvi* (Prague), 1984, no. 3, reports that Czechoslovakia consumes between fifty and eighty percent more energy than does Austria, France, or Japan to produce similar units of national income. Cited in Radio Free Europe, *Situation Report* (hereafter RFE-*SR*), no. 17 (Czechoslovakia), 21 September 1984. Kramer, *The Energy Gap in Eastern Europe,* pp. 110–13, examines the variety of factors contributing to the excessive consumption of energy.

4. *Pravda* (Moscow), 23 September 1990.

5. Unless otherwise noted, the following analysis is drawn from John M. Kramer, "Soviet-CMEA Energy Ties," *Problems of Communism,* July/August 1985.

6. See, for example, *Voprosy ekonomiki* (Moscow), 1971, no. 12. *New York Times,* 29 November 1969, reported that an official in Czechoslovakia said his country would soon get less oil from the USSR because "Soviet output would be shifting to Siberia."

7. Cited in Radio Free Europe, *Background Report* (hereafter RFE-*BR*), 24 August 1984.

8. A Hungarian source reports that as early as the sixties, the Soviet Union asserted that it would accept only hard goods for much of the energy it sold to Eastern Europe (*Kozgazdasagi Szemle* [Budapest], November 1979, cited in RFE-*SR*, no. 22 [Hungary], 5 December 1979).

9. For a detailed examination of these projects, see L. Csaba, "Joint Investments and Mutual Advantages in the CMEA: Retrospection and Prognosis," *Soviet Studies,* April 1985. On this subject, also see J. B. Hannigan and C. H. McMillan, "Joint Investment in Resource Development: Sectoral Approaches to Socialist Integration," in U.S. Congress, Joint Economic Committee, *East European Economic Assessment,* Part 2 (Washington: U.S. Government Printing Office, 1981).

10. For a typical example of such criticism, see the interview with now retired deputy prime minister Joszef Marjai of Hungary as carried by Magyar Tavirati Irada (Budapest), 21 October 1988, in FBIS-*EEDR,* 24 October 1988, p. 32.

11. Honecker's remarks appear in *Neues Deutschland* (East Berlin), 12 April 1981. *Ekonomicheskaia gazeta* (Moscow), 1981, no. 45, reports the statement of the Comecon official.

12. *Wall Street Journal,* 16 March 1983.

13. See John M. Kramer, "Council for Mutual Economic Assistance," in Richard Staar, ed., *1988 Yearbook on International Communist Affairs* (Stanford: Hoover Institution Press, 1988), pp. 371–74, for an elaboration of the Gorbachev program.

14. *Pravda* (Moscow), 7 July 1988.

15. Moscow Domestic Service, 6 July 1988, in Foreign Broadcast Information Service, *Soviet Union Daily Report* (hereafter FBIS-*SUDR*), 12 July 1988, p. 3.

16. Data on trade turnover for 1987 from *Vneshniaia torgovlia SSSR v 1987* (Moscow: Financy i Statistika, 1988), table 3. Respective data for 1988 and 1989 from *Ekonomika i zhizn* (Moscow), 1990, no. 5, in FBIS-*SUDR,* 20 April 1990,

pp. 5–8. In 1989, compared to 1988, the USSR experienced a decline in the value of bilateral trade with Czechoslovakia, Hungary, and Poland. It registered marginal increases in this trade with Bulgaria and Romania.

17. The Soviet economist is a research associate at the USSR Academy of Sciences. His estimate of what the USSR lost in hard currency in 1988 by exporting oil to Eastern Europe appears in *Trud* (Moscow), 12 March 1990. The Western estimate of the price of Soviet oil to Eastern Europe in 1989 is from Radio Free Europe, *Report on Eastern Europe* (hereafter RFE-*REE*), 31 August 1990, p. 42.

18. As quoted by TASS (Moscow), 31 May 1990, in FBIS-*SUDR*, 31 May 1990, p. 24.

19. Prime Minister Karoly Grosz of Hungary reported in 1988 that the USSR was "ready to maintain existing ratios" in exports of energy to Hungary through 1995 (*Nepszabadsag* [Budapest], 9 July 1988). For the expression of similar pledges to Czechoslovakia and Poland, see, respectively, Prague Television Service, 17 July 1990, in FBIS-*EEDR*, 18 July 1990, p. 15; *Rzeczpospolita* (Warsaw), 14 September 1990, in FBIS-*EEDR*, 24 September 1990, p. 56. A Polish source reports that the projected volume of Soviet oil deliveries to Poland in 1991–95 was so contentious that it could be resolved only by the respective prime ministers of the two countries (*Tygodnik kulturalny* [Warsaw], 4 September 1988, in Joint Publications Research Service, *East Europe Report* [hereafter JPRS-*EER*], 23 November 1988, p. 2).

20. E.g., when Polish premier Tadeusz Mazowiecki journeyed to Moscow in December 1989, Soviet premier Ryzhkov reportedly "agreed to analyze the possibilities" of meeting Mazowiecki's request for increased deliveries of natural gas to Poland in 1991–95 (Warsaw Television Service, 5 December 1989, in FBIS-*EEDR*, 7 December 1989, p. 85). For a pledge to increase Soviet deliveries of natural gas to Bulgaria in 1990, see *Krasnaia zvezda* (Moscow), 17 April 1990, in FBIS-*SUDR*, 26 April 1990, p. 2.

21. On shortfalls in Soviet deliveries of crude oil in the first half of 1990, see, *inter alia*, Warsaw Domestic Service, 29 August 1990, in FBIS-*EEDR*, 29 August 1990, p. 46; *Rude pravo* (Prague), 15 February 1990; and Magyar Tavirati Irada, 10 July 1990, in FBIS-*EEDR*, 11 July 1990, p. 35. Premier Ryzhkov's announcement of the seven MT (million metric ton) cut in crude oil exports is carried by Moscow Television Service, 7 July 1990, in FBIS-*SUDR*, 9 July 1990, p. 36. *Komsomolskaia pravda* (Moscow), 29 August 1990, reports the prediction that the USSR will deliver 20 percent less crude to Eastern Europe in 1990 than the planned volume.

22. See, for example, the statements by Soviet officials reported in Ceskoslovenska Tiskova Kancelar (Prague), 19 January 1990, in FBIS-*EEDR*, 22 January 1990; and in Polska Agencja Prasowa (Warsaw), 19 July 1990, in FBIS-*EEDR*, 20 July 1990, p.

23. Deputy Premier Dlouhy's estimate is carried by Prague Television Service, 17 July 1990, in FBIS-*EEDR*, 18 July 1990, p. 15. Ceskoslovenska Tiskova Kancelar, 9 October 1990, in FBIS-*EEDR*, 12 October 1990, p. 22, reports the estimate by President Havel's press secretary of oil deliveries in 1990.

24. Osiatynski contends that a Soviet delegation, "acting on the instructions of its government," informed Poland of the termination of oil exports during talks in Moscow from 28–31 August 1990 (*Rzeczpospolita*, 14 September 1990, in FBIS-

EEDR, 24 September 1990, pp. 55–56). The Soviet ambassador to Poland dismisses as "rumors" that are "completely untrue" the intention of the USSR to end its oil exports to Poland (*Trybuna* [Warsaw], 13 September 1990, in FBIS-*EEDR,* 20 September 1990, p. 28). See Polska Agencja Prasowa, 17 October 1990, in FBIS-*EEDR,* 18 October 1990, p. 28.

25. Magyar Tavirati Irada, 11 September 1990, in FBIS-*EEDR,* 13 September 1990, p. 29; ibid., 5 September 1990, in FBIS-*EEDR,* 6 September 1990. Another source quotes Hungarian experts as stating that it is *"almost* impossible" to believe that the USSR would terminate oil exports to Eastern Europe in 1991 (ibid., 11 September 1990, in FBIS-*EEDR,* 13 September 1990, p. 29 [emphasis added]).

26. *Mlada fronta dnes,* 2 October 1990, in FBIS-*EEDR,* 10 October 1990, pp. 20–21.

27. TASS, 12 October 1990, in FBIS-*SUDR,* 15 October 1990, p. 20, reports on the deal between Czechoslovakia and Tyumen Oblast for oil in 1990. Another source reports that Czechoslovakia will pay for 10 percent of this oil in hard currency and obtain the rest for barter (*Zemedelske noviny* [Prague], 6 October 1990, in FBIS-*EEDR,* 15 October 1990, p. 26). The estimate that Czechoslovakia may in the future receive as much as three MTs of oil annually under similar arrangements is carried by Ceskoslovenska Tiskova Kancelar, 19 September 1990, in FBIS-*EEDR,* 20 September 1990, p. 11.

28. On these plans, see for Czechoslovakia, *Pravda* (Bratislava), 20 February 1989, in FBIS-*EEDR,* 27 February 1989, p. 20; for Hungary, Magyar Tavirati Irada, 13 February 1990, in FBIS-*EEDR,* 14 February 1990, p. 42; for Poland, Polska Agencja Prasowa, 25 November 1989, in FBIS-*SUDR,* 27 November 1989, p. 23. See John M. Kramer, "Energy and the Environment in Eastern Europe," in Joan DeBardeleben, ed., *East European Environmental Problems and Policies* (Montreal: McGill University Press, forthcoming) for a discussion of initiatives to mitigate environmental degradation through greater use of natural gas.

29. Polska Agencja Prasowa, 17 October 1990, in FBIS-*EEDR,* 18 October 1990, p. 28.

30. Magyar Tavirati Irada, 13 February 1990, in FBIS-*EEDR,* 14 February 1990, p. 42. On this subject also see Magyar Tavirati Irada, 18 June 1990, in FBIS-*EEDR,* 11 June 1990, p. 59.

31. *Pravda* (Moscow), 16 March 1990. The USSR has either closed or canceled the construction of fifteen nuclear power stations since the explosion at Chernobyl'. The Ukrainian SSR, where 40 percent of Soviet nuclear capacities are located, recently declared a five-year moratorium on building nuclear power plants. For an overview of the cutbacks in the Soviet nuclear program, see *Eastern Europe Newsletter,* 24 September 1990, pp. 3–4.

32. *Magyar Nemzet* (Budapest), 16 June 1990, in JPRS-*EER,* 12 July 1990. Hungary derives almost 30 percent of its indigenous consumption of electricity from imports from the USSR. The respective figure for all other East European states is less than 10 percent (*Pravda* [Moscow], 16 March 1990).

33. The provision is contained in a presidential decree on foreign trade. For details of the decree, see Moscow Domestic Service, 24 July 1990, in FBIS-*SUDR,* 30 July 1990, p. 35.

34. The Soviet criticism of the transferable ruble and system of barter trade is from *Izvestiia* (Moscow), 2 July 1990. For a typical East European criticism of

Comecon trading practices, which argues that they derive from "pre-Phoenician times," see *Tygodnik Solidarnosc* (Warsaw), 6 July 1990, in JPRS-*EER*, 16 August 1990, p. 40.

35. Polska Agencja Prasowa, 30 July 1990, in FBIS-*SUDR*, 31 July 1990, p. 18.

36. Throughout this study, I use $30 per barrel as the prevailing world-market price for oil. This is necessarily an approximation of the actual price at the moment, given the volatility of the current international oil market. For an analysis that argues that the world-market price for oil likely will be "settling in for an extended time" at more than $30 per barrel, see the *Washington Post*, 20 September 1990, p. E1.

37. The Soviet estimate on prices for machinery in Comecon trade is from *Izvestiia*, 8 May 1990. In contrast, a recent Western study, drawing on calculations by Hungarian economists, finds that "there is little basis for assuming" that exports of machinery and equipment now are overpriced in Comecon trade, although they may have been overpriced in the sixties (Kazimierz Poznanski, "Opportunity Cost in Soviet Trade with Eastern Europe: Discussion of Methodology and New Evidence," *Soviet Studies*, April 1988, p. 292).

38. *Eastern Europe Newsletter*, 24 September 1990, p. 5, provides the assessment of the European Economic Community's response to Prime Minister Antall's proposal. See Radio Free Europe, *Report on Eastern Europe*, 31 August 1990, for details of the Czechoslovak proposal and the assessment that the European Economic Community has rejected it.

39. For a Soviet analysis of the sundry opposition expressed in Eastern Europe to an immediate switch to hard-currency trade at world-market prices, see *Izvestiia*, 2 July 1990. East European sources advocating a phased introduction of the new policy include *Mlada fronta dnes*, 15 February 1990, in FBIS-*EEDR*, 21 February 1990, p. 8; *Nepszabadsag*, 10 August 1990, in FBIS-*EEDR*, 15 August 1990; and Polska Agencja Prasowa, 9 April 1990, in FBIS-*EEDR*, 16 April 1990, p. 52. Magyar Tavirati Irada, 11 July 1990, in FBIS-*EEDR*, 12 July 1990, p. 34, carries the argument that the USSR should shoulder part of the burden of mitigating the legacy of its rule in the region.

40. See *Izvestiia*, 8 May 1990, for details of the negotiations between Hungary and the USSR. The Czechoslovak source reporting on this subject is Ceskoslovenska Tiskova Kancelar, 4 October 1990, in FBIS-*EEDR*, 9 October 1990, p. 2.

41. Moscow Television Service, 26 July 1990, in FBIS-*SUDR*, 27 July 1990, p. 50. Katushev's remarks, and the other evidence presented in this paragraph, contradict reports in the Western press that beginning in 1991 the USSR will only export energy for hard currency to Eastern Europe. An example of such a report appears in the *Washington Post*, 12 August 1990, p. A25.

42. For details of the agreement, see *Izvestiia*, 5 September 1990. Deputy Premier Vales's remarks are reported by Prague Domestic Service, 1 September 1990, in FBIS-*EEDR*, 7 September 1990, p. 23. The agreement also provides in principle for the conversion into dollars of the USSR's ruble debt to Czechoslovakia. The same source reports Vales's comment that in 1991 this debt will be the "only or main source" whereby Czechoslovakia will pay for Soviet imports "being sold on the world market for cash—i.e., raw materials, crude oil." However, Vales admitted that the two sides still disagreed on the key question of the exchange rate to be used in the ruble/dollar conversion.

43. Commenting upon this agreement, a Hungarian source explained: "Both countries have to compile a list of products by September, stating which products they would require and what they would have to offer in return . . . the two parties are thus putting down on paper what goods they will be buying from each other in the future" (*Nepszabadsag*, 10 August 1990).

44. E.g., a Soviet source argues: "If reciprocal deliveries are retained . . . what would be the point of introducing foreign currency settlements in trade? After all, the dollar will then have to perform the present functions of the transferable ruble, in other words, it would have to serve as the unit for calculating the volume of commodity cross-flows, which are more or less equal. This will be no market" (*Sovetskaia Rossiia* [Moscow], 11 May 1990). A Hungarian source makes the same point more succinctly: "In my mind it is still not clear what difference this new system makes" (*Figyelo* [Budapest], 10 May 1990, in JPRS-*EER*, 91 [25 July 1990], p. 29).

45. The deterioration of $1.5 billion in Hungary's terms of trade with the USSR is carried by, among others, TASS, 31 May 1990, in FBIS-*SUDR*, 31 May 1990, p. 24. A Hungarian source accepting this estimate is reported by Magyar Tavirati Irada, 28 May 1990, in FBIS-*EEDR*, 29 May 1990, p. 28. See Magyar Tavirati Irada, 26 July 1990, in FBIS-*EEDR*, 27 July 1990, p. 16, for the estimate that Soviet oil will be three-and-a-half to four times more expensive for Hungary beginning in 1991.

46. *Zemedelske noviny*, 20 July 1990, in FBIS-*EEDR*, 26 July 1990, p. 11. Commenting on the impact of higher oil prices, another source contends: "It can be said without exaggeration that Czechoslovakia will be experiencing, with a delay of several years, the oil shock which hit the rest of the world twice in the seventies. It is highly questionable, however, whether we will be able to come to terms with it just as successfully as the advanced countries did" (*Rude pravo*, 19 July 1990, in FBIS-*EEDR*, 121 [22 July 1990], p. 20).

47. For example, Czechoslovakia has paid hard currency for one MT of Soviet crude in 1990 that it had "planned" to pay for in barter (*Pravda* [Bratislava], 5 September 1990, in FBIS-*EEDR*, no. 178 [13 September 1990], p. 13). Similarly, Poland has used "freely convertible currency" to purchase one MT of crude from the USSR in 1990 (*Trybuna*, 13 September 1990, in FBIS-*EEDR*, 183 [20 September 1990], p. 28). *Eastern Europe Newsletter*, no. 8 October 1990), p. 3, estimates that in 1991 Czechoslovakia will pay with hard currency for approximately one-half of the oil that it imports from the USSR.

48. As USSR minister of foreign economic relations Katushev observed, this provision will "pressure" the Eastern Europeans to improve the quality of their exports to the Soviet Union (Moscow Television Service, 26 July 1990, in FBIS-*SUDR*, no. 145 [27 July 1990], p. 50).

49. *Trud*, 12 March 1990. The critic is V. Voloshin.

50. For a commentary detailing these problems, see *Pravda* (Moscow), 27 June 1990.

51. Data on oil output in 1989 and 1990 from *Izvestiia*, 20 July 1990, and *Komsomolskaia pravda*, 29 August 1990. Soviet estimates of oil output in 1990 are reported by Radio Moscow, 24 August 1990, in FBIS-*SUDR*, 27 August 1990, p. 65; *Moscow News*, 1990, no. 43, p. 13. For estimates that production could fall below 500 MTs, see *Eastern European Newsletter*, no. 8 (24 September 1990), p. 8; Tanjug

Domestic Service (Belgrade), 8 October 1990, in FBIS-*EEDR*, 9 October 1990, p. 76.

52. *Izvestiia*, 20 July 1990.

53. Premier Ryzhkov examined these problems in a recent speech reported by *Pravda* (Moscow), 30 March 1990. Also see ibid., 4 August 1990, for another candid assessment of the problems besetting the oil industry.

54. *Washington Post*, 15 August 1990, p. D1, provides details of these agreements. It notes that the agreements with Texaco and Chevron are both "preliminary and highly conditional." Henry Shuler, an oil analyst at the Center for Strategic and International Studies, argues that at best the USSR can only hope "in short to medium terms to slow the rate of decline" in oil production through Western financial and technological assistance (ibid., 5 September 1990, p. F5).

55. Plan Econ, Inc. *Plan Econ Report*, 15 January 1988, p. 4.

56. *Ekonomika i zhizn*, no. 15 (April 1990).

57. As reported in Radio Liberty, *Report on the USSR*, no. 31 (3 August 1990), p. 30.

58. *Pravda* (Moscow), 23 September 1990.

59. As one Soviet source argues, "In conditions of drastic economic reform and genuine economic accountability all commodities, including fuel, raw materials, and power, must be sold at real world market prices. There can be no more mutual concessions on matters of quality and regularity of supplies" (*Izvestiia*, 10 January 1990).

60. *Trud*, 12 March 1990.

61. For Soviet expressions of such sentiments, see, *inter alia*, Bulgarsko Telegrafna Agentsiia (Sofia), 30 July 1990, in FBIS-*EEDR*, 31 July 1990, p 8; Prague Television Service, 19 July 1990, in FBIS-*SUDR*, 23 July 1990, p. 29. East European sources concurring with this assessment include *Zemedelske noviny*, 20 July 1990, in FBIS-*EEDR*, 26 July 1990, p. 11; Ceskoslovenska Tiskova Kancelar, 19 July 1990, in FBIS-*EEDR*, 19 July 1990, p. 12; *Rude pravo*, 19 July 1990, in FBIS-*EEDR*, 26 July 1990, p. 13. In contrast, the East Europeans could hardly find reassuring the advice of one Soviet official that they take "into account" that energy exports can be used as a "political weapon" against them (quoted in the *Washington Post*, 10 November 1989, p. A41). It is impossible to determine from publicly available materials to what extent this warning reflects the thinking of the Soviet leadership on this subject.

62. A Soviet source explains how the former Communist regimes of Eastern Europe raised the specter of political instability to extract more oil from the USSR: "All a leader from one of the fraternal countries had to do was to bamboozle one of the higher representatives of the Soviet leadership with the possibility of political instability . . . and he then could boldly appeal for Soviet assistance, including Soviet petrodollars" (*Pravda* [Moscow], 23 September 1990).

63. One Soviet critic chastised the Eastern Europeans for holding a "parasitical attitude towards the Soviet Union" by engaging in such trade (*Trud*, 12 March 1990). Data on Hungary from Magyar Tavirati Irada, 26 July 1990, in FBIS-*EEDR*, 27 July 1990, p. 16. See Kramer, *The Energy Gap in Eastern Europe*, pp. 104–5, for a discussion of the complex and often imaginative ways that East European states participated in the reexport trade.

64. Ceskoslovenska Tiskova Kancelar, 7 October 1990, in FBIS-*EEDR*, 11 October 1990, p. 15.

65. For example, responding to entreaties from Prime Minister Andrei Lukanov of Bulgaria that the USSR fulfill its oil contract with that country, Soviet premier Ryzhkov "made it clear that the problem was related to Bulgaria's deliveries to the USSR. He pointed out that there were products of technical designation the USSR could not do without. He laid stress on Soviet requirements of foodstuffs and industrial consumer goods. He reminded him of the cigarette-shortage crisis of the past months, hinting that it had been aggravated not only through the failures of the local producers but also by 'our friends failing to effect deliveries' " (Bulgarsko Telegrafna Agentsiia, 25 September 1990, in FBIS-*SUDR*, 188 [27 September 1990], p. 29).

66. *Sovetskaia Rossiia*, 11 May 1990.

67. Premier Calfa of Czechoslovakia estimates that these factors make oil imported from the Middle East upward of 30 percent more expensive than oil imported from the USSR (*Mlada fronta dnes*, 8 October 1990, in FBIS-*EEDR*, 199 [15 October 1990], p. 25).

68. *Washington Post*, 5 September 1990, p. F1. This estimate assumes that oil prices continue to "hover" at $25 per barrel for the remainder of 1990. In actuality, world-market prices for oil during this period may average over $30 per barrel, thereby bringing the USSR even greater revenue from oil exports.

69. Quoted in ibid. For a Soviet source that shares this assessment, see *Rabochaia tribuna* (Moscow), 12 August 1990, in FBIS-*SUDR*, 20 August 1990, p. 66. In contrast, officials in Czechoslovakia, Hungary, and Poland fear that the USSR might silently sell for hard currency "on the international market oil intended for them." Reportedly, these states may seek to pressure industrial nations against purchasing this oil "although maneuvering scope in this respect is rather narrow" (*Hospodarske noviny* [Prague], 4 October 1990, in FBIS-*EEDR*, 195 [9 October 1990], p. 2).

70. Reports detailing the deleterious consequences of the decline in Soviet oil deliveries include Bulgarsko Telegrafna Agentsiia, 11 July 1990, in FBIS-*EEDR*, 12 July 1990; Warsaw Domestic Service, 14 July 1990, in FBIS-*EEDR*, 16 July 1990; *Pravda* (Bratislava), 5 September 1990, in FBIS-*EEDR*, 13 September 1990, p. 14. Hungary has experienced the most strident public protests against recently enacted increases in prices for fuels and power. Protesting against higher prices for gasoline, taxi drivers in Budapest and several other cities paralyzed traffic by blockading bridges and other heavily traveled arteries. For details of these protests, see Budapest Domestic Service, 26 October 1990, in FBIS-*EEDR*, 26 October 1990, p. 25. Similarly, an "explosive" situation arose in Czechoslovakia when the government enacted price increases for energy to compensate for less Soviet oil (Ceskoslovenska Tiskova Kancelar, 19 July 1990, in FBIS-*EEDR*, no. 140 [20 July 1990], p. 13).

71. *Eastern Europe Newsletter*, no. 8 (October 1990), p. 1.

72. Prague Television Service, 3 October 1990, in FBIS-*EEDR*, 4 October 1990, p. 14. *Hospodarske noviny*, 4 October 1990, in FBIS-*EEDR*, 9 October 1990, p. 2, reports on the possibility of a joint approach by Czechoslovakia, Hungary, and Poland for more oil from the USSR.

73. Washington Post, 28 September 1990, p. A21.

74. Overall, Iraq owes an estimated $4 billion to Eastern Europe. Estimated debt to individual states include Romania ($1.7 billion), Bulgaria ($1.2 billion),

Czechoslovakia and Poland (each $500 million), and Hungary ($145 million) (Radio Free Europe, *Report on Eastern Europe*, no. 34 [24 August 1990], p. 42).

75. Data for Bulgaria from Bulgarsko Telegrafna Agentsiia, 11 August 1990, in FBIS-*EEDR*, 13 August 1990, p. 2; for Poland from Polska Agencja Prasowa, 23 August 1990, in FBIS-*EEDR*, 24 August 1990, p. 18; for Hungary from Magyar Tavirati Irada, 6 August 1990, in FBIS-*EEDR*, 7 August 1990, p. 13; for Czechoslovakia from *Verejnost* (Bratislava), 9 August 1990, in FBIS-*EEDR*, 13 August 1990, p. 3.

76. Moscow World Service, 28 August 1990, in FBIS-*SUDR*, 29 August 1990, p. 4.

77. Information on Italy and Venezuela from, respectively, Ceskoslovenska Tiskova Kancelar, 5 October 1990, in FBIS-*EEDR*, 6 October 1990, p. 18; *Hospodarske noviny*, 28 September 1990, in FBIS-*EEDR*, 4 October 1990, p. 13.

78. Magyar Tavirati Irada, 3 September 1990, in FBIS-*EEDR*, 4 September 1990, p. 25.

79. Data for Bulgaria from Bulgarsko Telegrafna Agentsiia, 21 September 1990, in FBIS-*EEDR*, 24 September 1990, p. 21; for Czechoslovakia from Prague Television Service, 12 October 1990, in FBIS-*EEDR*, 15 October 1990, p. 25; Prague Domestic Service, 10 October 1990, in FBIS-*EEDR*, 12 October 1990, p. 23; for Poland from Polska Agencja Prasowa, 4 October 1990, in FBIS-*EEDR*, 5 October 1990, p. 35.

80. "Both the World Bank and the International Monetary Fund expect to pay special attention to the group of East European countries caught in the economic crunch between Iraq and the UN embargo," according to the *Washington Post*, 25 September 1990, p. D7. East European sources that discuss how these institutions may assist Eastern Europe to mitigate the effects of the Gulf crisis include Prague Domestic Service, 28 September 1990, in FBIS-*EEDR*, 28 September 1990, p. 22; Budapest Domestic Service, 28 September 1990, in FBIS-*EEDR*, 28 September 1990, p. 30. Neither institution has yet announced the actual amount of assistance Eastern Europe will receive to this end.

81. Budapest Domestic Service, 27 September 1990, in FBIS-*EEDR*, 28 September 1990, p. 29.

82. For a detailed discussion of the need for energy conservation in Eastern Europe, see Kramer, *The Energy Gap in Eastern Europe*, pp. 109–25. Unless otherwise noted, materials on energy conservation herein are derived from this source.

83. *Komsomolskaia pravda*, 29 August 1990.

84. See ibid. for a discussion by a Soviet source of these measures. For details of the price increases, see for Czechoslovakia, *Pravda* (Bratislava), 12 April 1990, in FBIS-*EEDR*, 17 April 1990, p. 19; for Hungary, Budapest Domestic Service, 29 June 1990, in FBIS-*EEDR*, 29 June 1990, p. 34; for Poland, Polska Agencja Prasowa, 30 June 1990, in FBIS-*EEDR*, 3 July 1990, p. 51.

85. Heti Villaggazdasag (Budapest), 11 October 1986, in JPRS-*EER*, 6 March 1987, p. 3.

86. For a detailed discussion of the relationship between excessive consumption of energy and environmental degradation in Eastern Europe, see Kramer, "Energy and the Environment in Eastern Europe."

Ecology and Technology in the USSR

Joan DeBardeleben

Ever since Stalin's "revolutions from above" in the thirties, the Communist party of the Soviet Union has presented itself to the population as the purveyor of technological progress. In the Brezhnev period, the ideological slogan "the scientific-technical revolution" proclaimed the capability of the Soviet socialist system to maximize the benefits of scientific and technological progress to realize its utopian social program. At the same time, in that period, the backwardness of Soviet technology and systemic obstacles to technological innovation became increasingly evident not only to the leadership but also to broader segments of the population. Furthermore, areas where technological advances seemed most credible, in the space program and in the military sector, hardly brought direct benefits to the population. In the mid-sixties, other doubts about the benefits of proclaimed technological progress also surfaced: the application of technology in both the industrial and agricultural sectors was directed toward the realization of higher volumes of material output. Not only was quality insufficient but, at the same time, the application of technological processes contributed to the degradation of the quality of life through its impact on the natural environment. While these effects have had little influence on the practice of technology development, they marked the beginnings of awareness in the USSR that the environmental crisis placed new demands on that sector.

The author is grateful to McGill University for support to facilitate this research and to John Kramer for comments on an earlier draft of the paper.

Ecology and Technology Before Perestroika

In the pre-Gorbachev period, the actual impact of ecological considerations on technological development was minimal. At the same time, some natural scientists and economists (whom I call, loosely speaking, environmental advocates) began to factor ecological effects into their assessment of the social and economic value of technological innovation. Some of their conceptions made their way into official parlance, most notably where they were consistent with other regime goals. For example, support for recycling of the by-products of production (in Soviet terminology, "the utilization of secondary raw materials") and other measures to reduce waste of natural resource inputs have been included in planning documents since the late seventies. Such commitments, consistent with leadership efforts to improve the efficiency of production and lower production costs, also implied support for transformation of the technological base of production. Nonetheless, most proposals made by experts to alter technology in response to ecological stresses did not make their way into practice. It is important to examine some of the theoretical underpinnings and policy considerations that explain this failure and also to examine more closely the case for technological conversion made by the environmental advocates.

Soviet Marxism, Ecology, and Technology

Underlying official Marxism-Leninism in the USSR was a strong technological optimism. This assumption was derived in part from Marxist theory itself, but was also rooted in the legitimatory structures that grew out of the Stalinist period. The sacrifices endured by the population during the rapid industrialization campaigns of the thirties were justified, in part, by the scientific foundation for the policy and its proclaimed economic and technological achievements. Technology was viewed as neutral or benign in itself, but fulfillment of its positive potential was seen as possible, and indeed almost inevitable, in the context of socialist productive relations characteristic of Soviet society. In this view, under socialism any damage to nature could be overcome by some other technological innovation; nature was to be mastered—and could be. The point was to "transform" nature (*preobrazovat' prirodu*), not to adapt human technology to the needs of nature. This goal seemed more plausible, given the vast and seemingly inexhaustible expanses of nature in the USSR.

Technological optimism was expressed in theoretical conceptions developed by leading Soviet scientists as well, most notably in the thought of the late geologist V. I. Vernadskii. For Vernadskii the noosphere represented the final stage in the evolution of the biosphere. "Man, taken as a whole, becomes a powerful geological force. And before him, before his thought and labor, stands the question of the rebuilding of the biosphere in the interests of freely thinking humanity as a unified whole."[1] Vernadskii's concept of the noosphere provided scientific justification for the mastery of nature in the name of scientific communism. In the sixties, the biologist G. F. Khil'mi elaborated a similar notion, the "biotechnosphere," which depicts a symbiosis of nature and technology based on "large-scale structures transforming the atmosphere, the hydrosphere, and the lithosphere of the Earth."[2] Both of these "scientific" concepts served ideological functions, as they legitimized the regime's technological optimism and its interventionist approach to nature.

The commitment to mastering nature underlay the nature-transforming projects characteristic of the Stalinist and post-Stalinist periods. These included, most prominently, the water-management and energy systems. Huge hydroelectric complexes had a profound impact on local ecosystems (for example, eventually transforming the Volga from a river to a series of dams, reservoirs, and canals). Under Brezhnev, land reclamation and irrigation projects took on vast dimensions, particularly in central Asia and in the non-chernozem region of Russia. Ecological concerns had very little, if any, impact on the manner in which such schemes were conceived and carried out. Only in some rare instances—for example, following public outcry over the damage to Lake Baikal—were there even minimal attempts to adjust technology to reduce its harmful impact on the environment. Furthermore, gigantomania was not limited to water management. Construction of huge metallurgical, chemical, and other manufacturing complexes mirrored the planners' heady confidence in their ability to manipulate material and nature to their own ends. Even in those cases where technological equipment was installed to reduce negative environmental effects (e.g., filters or scrubbers to detoxify emissions into air or water), it was not well maintained and often was not functioning at all.

Economic goals and structures, as well as the ideological factors discussed above, also mediated against an ecological assessment of technological choices. Environmental protection was a low rung on the ladder of regime priorities. Levels of material output were the primary concern. The low status of environmental concerns was reflected in the near absence of personnel trained in environmental amelioration and in related fields of research and technology development. Furthermore, incentive structures in the planning system itself did not encourage resource-saving techniques or recycling, let alone conversion to less polluting production technologies. Particular ministries and enterprises had almost no incentive to reduce externalities as they pursued their planned production quotas; since most natural resources were granted to them free of charge, they also had every motivation to squander water, land, or mineral resources, if such actions eased fulfillment of material output quotas.

The Critique: The Call to Ecologize Production

Despite formidable obstacles to change, beginning in the sixties scientists and economists in the USSR began to propose the main features of a new technological policy that would be responsive to environmental concerns. As they articulated their ideas, however, they were constrained by official limits on public debate. While these fledgling environmental advocates could argue for some forms of technological transformation, they had to do so within prescribed limits and could not publicly contradict certain official values. At least on an explicit level they, for the most part, cloaked their arguments in terms consistent with the prevailing norms of the system—that is, the commitment to optimal rates of economic growth, faith in new technology to generate solutions, the definitive advantages of a central planning system, and the superior capability of socialist society to realize ecological goals. At the same time a counterstrain of "ecological pessimism" pervaded some of these critiques, and one could begin to detect in the Soviet scholarly debate a subtle suggestion that technology itself is neither neutral nor benign.[3] Sometimes such ideas were smuggled into the debate by quoting Western authors whose ideas were explicitly disavowed, but whose views were actually more widely disseminated for having been quoted. Critics also emphasized that socialist society itself does not automatically make ecologically sound technological choices; rather, these must emerge through a process of ecological education, scientific research, and policy deliberation.

Environmental advocates coined their own slogan—"ecologization of production." According to this conception, human intervention in nature should serve to increase the productivity of nature, improve its capacity for self-regeneration, and reinforce its internally balanced biological cycles. The notion did not involve a retreat from human intervention, but rather an adaptation of technological methods to enhance and accommodate the needs of nature. Beyond that, ecologization of production implied that human society could learn from nature, for by studying its techniques of self-regeneration, society could try to reproduce them in technological processes. Human production would become a "technical imitation of nature."[4] This concept did not challenge the goal of economic growth in an explicit manner. But it also represented more than a simple effort to introduce ameliorative devices on already existing technology (emission controls, filters, scrubbers, etc.).

On a practical level, the most radical implication of ecologizing production involved a shift toward use of "closed-cycle production" and low-waste technology. In addition to the ecological payoff of such a strategy, economists emphasized other advantages, consistent with regime priorities. Such technology would increase the productivity of labor and stimulate more efficient use of material inputs; it would thus contribute to the much-touted goal of shifting the productive structure from an extensive strategy (the Stalinist approach based on expanding the range of material, natural, and labor inputs in the productive system) to an intensive strategy (getting more output from limited and increasingly scarce inputs). Critics of existing production patterns contrasted human productive systems with natural ones. Whereas natural processes are largely cyclical (involving the building up of new life forms and their subsequent decomposition), human productive processes have been largely unilinear, utilizing only a minuscule portion of materials extracted from nature and leaving the rest as unutilized, and often polluting, waste. The adoption of closed-cycle production was presented as particularly important because productive processes had become so intense and expansive that nature itself no longer had the capacity to break down wastes generated by society. Humans themselves would have to include organic and microbiological processes in their technology.

While this view gained numerous adherents in scholarly circles[5] (especially among philosophers, biologists, and economists), its effect on policy was strictly limited, for implementation of the new strategy would necessitate a fundamental and broad-ranging retooling of the Soviet productive structure. Some success was achieved in instituting closed-cycle water systems in new industrial facilities. Official statistics indicate that, in 1988, 72 percent of water consumption for production needs was provided by recycled or successively used water (up from 69 percent in 1985); they also indicate a loss of 18.7 percent for irrigation water during transport to the point of discharge.[6] These figures probably overestimate the efficiency of the facilities in conserving water.

A less radical corollary of the environmentalists' blueprint gained greater official support—namely, the notion of expanded use of "secondary raw materials" (recycling of productive waste). In 1978 bonuses were put in place to encourage the collection, storage, and shipping of scrap metal; and beginning in 1981, state, ministry, republic, and enterprise plans were to include a section on the utilization of secondary raw materials. Other policies to reduce waste of materials in the production process were also instituted in the eighties with some significant results in certain sectors of the economy. Specialists emphasized the economic as well as the ecological benefits of the recycling efforts. Major shifts in the technological basis of production did not, however, occur. Rather, most progress was made in those sectors where recycling was relatively easily accommodated to existing productive structures. Programs were also put in place to gather recyclable waste (e.g., paper and bottles) from the population at large, although the incentive structure was insufficient to bring maximum response.[7]

Ideological constraints affected the manner in which environmentalists could develop their notion of ecologizing production. Methods to achieve the goals were most often discussed in terms of large-scale production complexes that could realize complex patterns of waste reuse.[8] Such schemes were consistent with the system's normative commitment to centralized production structures, gigantic production systems, and supposed economies of scale. This approach was rooted in a particular conception of the "ecologizing of production": ecology involves interdependence ("everything is connected to everything"), which implies complexity and centralization; engineered closed-cycle cybernetic systems on a large scale were suggested to capture this complexity. This logic contrasts sharply with an approach that enjoys greater popularity in the West—the "small is beautiful" notion and concepts of "appropriate technology." This Western variant is based on a different reading of ecological demands: here ecological production seeks to imitate the self-regulation of natural processes, which then is taken to imply social self-management and local autonomy in the West. To realize this end, small-scale technology, geared to local needs, is seen as superior (composting, solar panels in homes, bicycle transport).

The preference for centralized ecological production expressed in Soviet writings of this pre-perestroika period had political roots (as did the "small is beautiful" idea in the West). Demands for local control over small-scale technology would have threatened central control and would, at least on the surface, have seemed difficult to implement within centralized economic structures and imperatives. The obstacles to small-scale ecological production were also rooted in the incentives that motivated enterprise behavior and in the general ideological environment. Indeed, local enterprises did often develop such "unplanned" local capacities to respond to bottlenecks in provision of inputs.[9] The problem was that this capability was not generally used for ecological ends, since there was no incentive to do so and ecological know-how was absent. Local initiatives at the enterprise level were rather geared to securing a stable base for the mandated material-output quotas in the state plan.

Thus, prior to perestroika, the theoretical groundwork had been laid by some scientists and economists for a shift in the technological orientation of production to realize a certain conception of ecological responsibility. Implementation of even this ideologically constrained strategy, however, confronted structural and economic roadblocks. Since 1985, some of these obstructions were weakened, but new problems have emerged to replace them.

Technology and Ecology under Gorbachev

Since the advent of perestroika and glasnost, possibilities for technological transformation in response to ecological demands have increased; several factors, discussed below, could, at least in theory, facilitate the adoption of more environmentally friendly technology. However the general political crisis and economic chaos have left few investment funds for technological innovation. Furthermore, while public attention was focused on ecological problems in the early years

of perestroika—especially following the Chernobyl nuclear-power accident in April 1986—since 1990 the larger economic and political
crisis has lowered the relative salience of ecological demands and,
thus, reduced pressure for technological conversion.

Chernobyl, Technological Pessimism, and Public Protest

The accident at the Chernobyl nuclear power plant in 1986 represented
a watershed in public attitudes toward technology, especially high
technology. Not only did numerous grass-roots movements grow up in
opposition to planned or existing nuclear facilities throughout the
USSR but the catastrophe also instilled wide-ranging skepticism toward the orthodox view of technology as intrinsically benign. Official
analyses of the causes of the accident emphasized individual human
error, attempting to vindicate both the technology itself and the social
system in which it operated. However, following the accident a major
shift in nuclear-power technology occurred; Soviet authorities with-
drew plans to develop further reactors of the Chernobyl type (the
RBMK, graphite moderated channel-type reactor) in favor of the other
predominant model operative in the USSR (VVER, a water-moderated
reactor, similar to most Western models). This shift suggested that
even if the technology itself was not completely to blame for the accident, it was at least partly responsible for the failure. The implication
was, then, an indirect admission to the public that indeed some technologies are less benign than others. The RBMK reactor, previously
hailed as the "Soviet" type, was acknowledged, in practice, to have some
inherent weaknesses. This undermined traditional claims about the
uniquely positive potential for technological progress under socialism.

Not surprisingly, in the months and years following Chernobyl,
large parts of the public became vocal in questioning official reassur-
ances about ecological safety, a debate that was made possible by the
expansion of glasnost from 1986. Not only did nuclear power come
under attack but also other industrial processes were subject to public criticism, including those in the pharmaceutical industry, the
paper industry, biochemical production, synthetic fertilizer and pesticide production, as well as factors in traditional heavy industrial sectors such as metallurgy and the chemical industry. Protests against
various kinds of ecological damage sprang up all over the USSR and

led to the closing or cancellation of numerous production facilities in diverse sectors of the economy, along with the demand that plants be converted to less polluting production processes. Gigantic water-transformation projects (such as the Siberian river-diversion project and the Volga–Chograi canal) were canceled (at least temporarily), and gigantomania itself was attacked.[10] A crisis of sorts resulted because technological retooling of the existing large-scale facilities would require large capital investments, expertise in areas weakly developed in the USSR, and, in many cases, the importation of foreign equipment. The facilities that were shut down under public pressure could not be quickly retooled; therefore, these closures aggravated the chronic shortage conditions in the Soviet economy. V. Bushuev, chair of the USSR Supreme Soviet subcommittee on energy, reported in late December 1990 that "construction work has been suspended or operations have been shut down at seventy power plants with a total capacity of 150 million kilowatts, which is half the current capacity of the entire unified power system. At present, construction is not being started on a single new power plant."[11] Thus, until now, the impact of the ecological movement on technological change has been largely negative (bringing cancellations of plans or closures of numerous facilities operating with damaging technological processes) rather than positive (eliciting alternatives such as the installation of new "ecologically friendly" technological processes). In late September 1990 Gorbachev himself issued a decree mandating that delivery targets be fulfilled; some of the closed plants were to be reopened (especially in the pharmaceutical industry).

Decentralization

Environmental demands were among the first to be articulated by the popular-front movements that emerged in the Baltic republics in the late eighties, and this pattern has been replicated in other republics, including parts of Russia. These grievances have been linked to more general claims for republic or regional self-management and for local decision-making power in the economic/ecological sphere. The pressure for regional economic control represents a rejection of the power of the large centralized ministries and of their right to mandate the construction and mode of operation of productive facilities in particular localities. In practice, the collapse of central state authority that ensued during the "parade of sovereignties" in 1990 in the USSR further undermined the capacity of central economic organs to control the course of economic development in the various regions in the USSR.

While bureaucratic wrangling leaves the fate of some projects that were canceled under popular environmental pressure at least theoretically open, a major shift in attitude has occurred. Big is no longer necessarily beautiful, for "big" is closely associated with the large central economic ministries, which have imposed their will on the various localities. Like the earlier official commitment to large engineering complexes, the new public skepticism toward these same projects is also politically based. For as local populations desire more control over their destinies, they object more strongly to large projects that necessarily imply centrally controlled investment, influxes of laborers from the outside, and massive intrusions into the local environment.

Therefore, local and regional ecological movements are now more sympathetic to the alternative ideology of "small is beautiful," with its emphasis on local control and self-regulation. Reinforcing this proclivity is the fact that some small-scale initiatives can proceed with more modest immediate investments (for example, reductions in the use of chemical fertilizers and pesticides). At the same time, assertion of local control over economic development requires that local authorities respond to numerous conflicting pressures, including demands for better goods and services. For example, the desire for energy self-sufficiency in Lithuania may make it difficult to forego the contribution of the otherwise unpopular Ignalina nuclear power station (which operates with the Chernobyl-type RBMK reactors, which is largely staffed by Russian skilled laborers, and which reflects the continuing influence of the central ministerial structure). Local autonomy may well require developmental choices, based on available natural resources, which will have negative impacts on the local environment or on ecosystems of neighboring regions. For instance, support for the Siberian river-diversion project continues to be expressed by some Central Asian elites, despite the presumed deleterious effects that project might have on the larger Eurasian ecosystem and particularly on the north. Furthermore, the start-up costs of developing some new technological options (e.g., solar power in Central Asia) may be too great for the still weak republic and local governments to bear; likewise, technological retooling of large enterprises and the cleanup operations relating to past production involve immense costs which specific localities might not be able to bear. To address these problems, local governments would certainly need the power to extract funds from the same local industries that have caused the pollution and would most likely require central subsidies as well.

Environmental critiques emanating in Lithuania suggest some possible avenues of regional ecological restructuring. One suggestion is that obstacles to ecological innovation that result from the highly bureaucratized and centralized ministerial systems could perhaps be overcome if regional authorities were given the power to make decisions on technological restructuring. Another proposal involves technological alterations at the Mazeikiai oil refinery to increase its production of unleaded gasoline, which would be exported beyond Lithuania's borders only after personal and public-transport needs at the local level had been met, thus assuring maximal positive benefit to the local ecosystem and population.[12] Such control over local distribution of environmentally friendly production would provide the population and the republic's government with an incentive to fund somewhat costly processes of technological conversion.

Other proposals, emanating from the Lithuanian Green movement,[13] have involved technological innovations such as the development of an instrumentation network and program to monitor radioactive emissions from Ignalina, the reequipping of the Akmencementas cement factory (in Naujoji Akmene in northern Lithuania), the reduced use of chemical pesticides and termination of aerial spraying, local control over storage and distribution of chemical fertilizers, and improvement of safety features and cleaning devices for equipment at the Kedainiai chemical plant and the Jonava Azotas factory. While such demands may seem to differ little from a wide range of causes that environmental advocates championed in official media outlets even before 1985, it is their local and specific focus that is significant. Particular technological goals (rather than aggregate plan targets) are being discussed, and the decentralization of economic power would eliminate numerous obstacles to their actual implementation. Elected local authorities are not only accountable to central state organs or Party bodies, which are more likely to enforce old priorities and to allow innovative approaches to become bogged down by departmental self-interest, but are also, at least in some sense, accountable to a public constituency. This fact changes the incentive structure for public-policy formulation. Indeed, already in some cases referenda are planned or have been held to allow expression of public opinion on developments with contested ecological effects.[14] Unfortunately, such expressions are usually "yea" or "nay" propositions and make it easy for the public to mandate plant closings without considering the opportunity costs and prospects for the resultant filling of production gaps. Furthermore, accountability of local authorities to the public can be meaningful only if the local government actually has the economic capacity to alter local development priorities and, thus, technological choices. This power, at least in most regions of the USSR, does not yet exist.[15]

Economic Reform

Introduction of market mechanisms in the economy should, overall, have a positive impact on incentives for technological innovation.[16] So far, however, the reform has not progressed sufficiently to realize this advantage. Despite the generally positive prospects that market reform presents for technological innovation, the likely impact in the environmental sector is less clear. The initial capital investment for a shift to environmentally friendly technology may not be rewarded in a market environment. Rather, the market may more likely encourage the externalization of environmental costs, as it does in the West when specific state regulatory mechanisms are inadequate. Soviet legislation[17] envisages the introduction of certain regulatory mechanisms, such as charges for routine and excessive pollution and charges for natural-resources use, which would implant an incentive structure to make it economically advantageous for enterprises to act in an environmentally responsible manner. Measures that have been proposed, however, would more probably encourage simple ameliorative responses rather than more thoroughgoing processes of technological conversion. Only where resource saving would be substantial enough to justify initial investment costs would the market encourage such conversion to new technological (closed-cycle or low-waste) production processes. Integration of the Soviet economy into the larger world market might also, through both joint ventures and imperatives to meet world standards in exports, encourage some technological upgrading. However, by the same token, in an effort to make investment opportunities attractive to foreign partners, Soviet authorities may just as well be tempted to lower environmental standards, as has happened in many third-world countries. All of these factors suggest that the implications of market reform (questionable as its achievement appears in early 1991) may be more ambiguous in encouraging environmentally sound technology than it would be in realizing other goals of technological innovation.

Technological Conversion: Selected Issues

In this section, I will examine two specific aspects of production where pressures for technological conversion have grown in the past three decades, but since 1985 in a more open and widespread manner. These involve (1) agricultural technologies specifically related to the widespread use of chemical pesticides and fertilizers (2) and waste reduction in the industrial, energy, and extractive sectors.[18] Here I will try to clarify the nature of the environmental demands and prospects for their realization in light of the factors discussed in the previous section.

Chemical Fertilizers and Pesticides

Environmental advocates in the USSR generally acknowledge that chemical pesticides and fertilizers have been overused in the USSR, leading to adverse health effects on both agricultural workers and the local population, declines in soil productivity, and toxic residues in water and produce.[19] The latter have in turn been linked to elevated rates of illnesses and mortality, particularly in Moldova, Armenia, and Central Asia.[20] Careless aerial spraying represents another hazard. In addition, local environmental pollution resulting from production of chemical preparations has also been a cause for concern. Environmentalists in Lithuania claim that 82.4 percent of residents within six kilometers of the Jonava Azotas plant (a chemical plant whose output includes chemical fertilizers) suffer from eye and nasal-passage irritation. Furthermore, higher than normal levels of respiratory disease, angina, and conjunctivitis characterize the region. Bronchitis is reportedly nine times above the republic's average for children in the Jonava region. An accident at the Jonava facility on March 20, 1989, occurred when a reservoir tank containing ammonium leaked and exploded. Toxic gases escaped, local residents were evacuated, and several hundred people reportedly suffered ill effects.[21]

Despite official recognition of the problem of overuse of chemical preparations in agriculture, plan targets for their production and application continued to rise in the late eighties. Bitofus, a pesticide used in cotton production, was banned in the late eighties, owing to its high toxicity and adverse health effects on the agricultural work force. On the other hand, residues of DDT exceed maximum permissible levels on more than 10,000 hectares of agricultural land;[22] some critics believe that DDT is still being applied in some areas, despite its prohibition. The report on the state of the environment issued by Goskompriroda (the newly created State Committee on Protection of Nature) indicates that in recent years use of chemical pesticides has declined. "In comparison with 1986, by 1988 the area of their application declined by 21 million hectares and the area using biological methods increased by 2.3 million hectares." Still, in 1988, only 26.7 million hectares saw use of biological methods compared to 165.7 million hectares of sown area utilizing chemical methods.[23] Research on microbiological processes of pest control is being pursued in various institutes in the USSR, but its translation into practice has been scattered.[24]

The statistical handbook on environmental issues and natural-resource use, issued for the first time in 1989, indicates that average consumption of mineral fertilizers per hectare of tilled land continues to rise in the USSR, although levels in 1987 were modest compared to some East European countries (Bulgaria, Hungary, the GDR, Czechoslovakia) and not excessive by Western standards.[25] (See Table 2, p. 166.) Recent criticism has centered not only on the volume of chemicals applied but also on their careless methods of storage and application, resulting in contamination of surrounding ecosystems and leakage of chemical components in the surrounding region, leading specifically to nitrate pollution of water resources. The application of organic fertilizers increased more slowly between 1980 and 1987 than did the application of mineral fertilizers. Thus, no technological shift is visible in this sector, despite vocal public criticism.

Republic authorities in several regions of the country have acknowledged the importance of improved control over and reduced use of chemical fertilizers and pesticides;[26] since 1980 the greatest advances in application of organic fertilizers have been achieved in Ukraine, Lithuania, Moldova, and Latvia. Obstacles include insufficient supplies of alternative preparations and continued pressures to fulfill plan mandates. The opening of independent cooperatives, privatization of agricultural land (enacted in the Russian republic in December 1990), the greater force of consumer preferences, and elimination of pressure to fulfill irrational plan quotas for chemical applications would all be factors encouraging technological conversion in this sector, should market reforms proceed. Public pressure, particularly in light of the evident health effects of overchemicalization, will likely continue, and public information about residue levels in food has expanded. Yet at this point, no major technological impact of the environmental movement is evident.

Waste Reduction

Reduction of productive waste has also proceeded relatively slowly in the USSR, although progress has been made in some areas in the past decade. The USSR's complex program for scientific-technical progress for the period 1991–2010 includes attention to resource-saving technology and waste reduction with its concomitant environmental advantages. Gorbachev himself announced in December 1990 that there would be an assertive effort in 1991 to enlist foreign capital to help upgrade technology in the raw-material and fuel and power sectors, and "in utilizing colossal amounts of waste products—with consideration for ecological requirements, of course."[27] Also in late December, the chair of the USSR Supreme Soviet subcommittee on energy reported losses of 30 percent in the energy sector. Based on estimates from the Siberian branch of the Academy of Sciences, he suggested that 17 billion rubles would be needed for gas scrubbers until the year 2000 to keep atmospheric emissions in the energy sector at present levels. Furthermore, "developing and adopting energy-saving technologies requires one-time capital investments of 50 to 100 percent greater than the cost of expanding traditional energy production."[28]

Goskompriroda's report cites several examples of achievements in this area, most notably in reducing the loss of raw materials in the process of extraction. In certain enterprises (e.g., some metallurgical and mining-enrichment combines), progress has been made in the complex use of mineral resources. Also, the extraction of sulfur from emission gases has improved in zinc- and copper-smelting factories. All of these technological improvements have reduced stress on the environment and have provided economic benefits.[29]

Inadequacies in waste reduction are even more striking than achievements, however. The most highly polluting industrial sectors

Table 1

Installations Put in Operation, by Economic Branch

Branch	Installations to capture and render harmless emission gases from stationary sources (1000s of M³ of gas/hour)			Installations to purify waste water (1000s of M³ daily)			Systems of water recycling (1000s of M³ daily)		
	1981–85 (average for year)	1987	1988	1981–85 (average for year)	1987	1988	1981–85 (average for year)	1987	1988
Energy-fuel complex	12,522	16,564	8,915	569	504	324	17,571	23,433	11,425
Metallurgical complex	15,592	21,633	8,742	413	158	1,125	2,561	1,750	1,675
Machine-building complex	3,405	1,850	1,806	522	394	346	285	769	217
Chemical-forestry complex	2,191	3,008	3,523	849	1,089	1,217	3,324	3,409	1,688

Source: Okhrana okruzhaiushchei sredy i ratsional'nye ispol'zovanie prirodnykh resursov v SSSR: statisticheskii sbornik (Moscow: Gosudarstvennyi komitet SSSR po statistike, Informatsionmo-izdatel'skii tsentr, 1989), pp. 151-52.

(energy, petrochemical, metallurgical, chemical) continued to underfulfill plan directions for utilization of productive wastes throughout the eighties and have failed to utilize all available capital-investment funds for this purpose. Installation of purification devices has also been consistently below planned levels in these sectors.[30] (See Table 1, p. 164.) Recycling efforts involving the consumption sector have also shown only slow improvement. Recycling of used paper showed only a minimal increase between 1985 and 1988 (and no improvement in terms of percent of volume of production of paper and cardboard). Relative to other East European countries, the USSR ranks last in paper recycling. In the late eighties, levels were approximately comparable to those achieved by the US.[31] (See Table 2, p. 166.) Use of other secondary raw materials follows a similar pattern, with only minimal gains in some areas (use of worn tires, polymeric secondary raw materials, wood waste, furnaces, scrap, and wastes of ferrous metallurgy), significant gains in the use of slag from blast-furnace production, and an actual decline in recycling of glass.[32] Other indicators of development of resource-saving technology show the USSR lagging substantially behind developed Western countries.[33] Statistics provided by Soviet sources generally suggest that the general economic crisis, which has been intensifying under conditions of perestroika, has had a neutral or somewhat negative impact on the capability of the system to reduce waste from the production process.

The weakness of the consumer sector in the USSR has had some beneficial side effects for the environment: less waste from throwaway goods; fewer automobiles and greater reliance on public transport; and less packaging. On the other hand, the shabby quality of many goods means that they do not have a long useful life; furthermore, shortage conditions encourage hoarding of goods, which may go unused or spoil before they find any useful life at all. Production of solid waste in the USSR still falls far behind that of most Western industrialized countries, although, as Table 2 indicates, methods of disposal rely more heavily on solid-waste dumps, with lesser use of incineration or composting.[34] Given the present economic crisis, it seems an academic question to ask about the ecological spin-offs of a transition to a consumer society in the USSR. On the other hand, there seems to be little discussion of the broader developmental/ecological issues at stake in forging a new economic-reform program. The prevailing goal, articulated both by the public and the leadership, is to increase production of

Table 2

Some Soviet Comparisons of Selected Data for Selected Countries

Country	Average consumption of mineral fertilizers [1]	Percent of paper and cardboard from recycled material (1987)
USSR	122	27
Czechoslovakia	311	32
GDR	367	50
Hungary	268	53
Poland	224	34
Canada	51	11
FRG	427	43
France	298	42
Great Britain	380	55
Italy	169	41
Japan	378	52
USA	106	26

Disposal of Solid Domestic Waste (SDW)[2]

Country	Vol. of SDW (million tons/year)	Estimate per capita[3]	Percent stored in dumps	Percent in-cinerated	Percent composted[4]
USSR	27.0[5]	.190[5]	97.0[5]	2.3[6]	0.75
Czechoslovakia	2.4	.144	89.5	8.0	2.5
GDR	4.5	.270	96.5	3.3	0.2
Canada	6.0	.236	80.0	19.0	1.0
FRG	28.0	.459	61.0	34.0	5.0
France	16.0	.291	46.4	41.0	12.0
Great Britain	16.5	.301	88.5	10.0	1.4
Italy	15.0	.263	67.0	18.0	10.0
Japan	32.0	.265	27.0	70.0	0.3
USA	235.0	.985	85.0	14.0	0.1

Sources: Gosudarstvennyi komitet SSSR po okhrane prirody, *Doklad: Sostoianie prirodnoi sredy v SSSR v 1988 gody* (Moscow, 1989), p. 77; *Okhrana okruzhaiushchei sredy i ratsional'nye ispol'zovanie prirodnykh resursov v SSSR: statisticheskii sbornik* (Moscow: Gosudarstvennyi komitet SSSR po statistike, Informatsionno-izdatel'skii tsentr, 1989), pp. 99–156.

Note: Sources of and methodologies for collecting data are not clarified in the Soviet sources.

[1] Average per hectare of tilled land.
[2] No year provided.
[3] Ton/year, based on estimated 1985 population.
[4] The three columns may not add up to 100% as in some countries other methods are also used.
[5] Data for RSFSR (Russian Republic) only.

Western-style goods. (Big Mac packaging in the USSR does at least have some redeeming value as a souvenir.) While improved public transport has official support, average Soviet citizens still aspire to an automobile culture. There has been little public reflection on the preferability of reusable over disposable products, probably because the latter do not, at the moment, seem a real option. In the search for sufficiency, the consumers' "green" instincts are understandably weak. Pressure for waste reduction in the consumer sector is likely to appear only in the far future. Thus, the technology underlying any transition to an economic mechanism more sensitive to consumer preferences may well overlook ecological concerns unless cost considerations, resource shortages, or a demand for better quality produce ecological benefits as unintended consequences.

Conclusion

While technology has had an immense and largely destructive impact on diverse ecosystems in the USSR, ecological considerations have not yet had a major influence on technological development. A major constraint inhibiting introduction of more ecologically sound technology is economic. Investment capital is simply not available from domestic resources, and foreign investors or partners are likely to resist avoidable costs to realize ecological goals, given the already considerable risks associated with doing business in the USSR. The one advantage the USSR may have is that so much of its capital equipment is outdated; its replacement, when it occurs, may allow Soviet firms to move to a new generation of technology more reflective of ecological concerns. This assumes that, in the interim, the USSR generates a research capability and trained personnel to develop new methods or that hard currency is available to buy them. These weak points in the ecological infrastructure pose as great an obstacle as the purely economic considerations, and here collaboration with Western research institutes and firms could be extremely important.

More likely, the old technology will stay in place for some time, despite public protests and criticism. The sheer scope of the problem will make it difficult for government leaders, planners, or enterprise directors to shut down or convert existing plants and processes at any but a most incremental pace. In some regions, such as the Baltic republics, the picture may appear a bit brighter because local initiative, the greater legitimacy of the local government, and the efficacy of popular pressure, along with an ability to attract international support, may facilitate an innovative search for local "small is beautiful" solutions.

Notes

1. M. N. Rutkevich and S. S. Shvarts, "Filosofskie problemy upravleniia biosferoi," *Voprosy filosofii,* 1971, no. 10, p. 62. This discussion of Vernadskii and Khil'mi draws heavily from my book *The Environment and Marxism-Leninism: The Soviet and East German Experience* (Boulder, CO: Westview Press, 1985), p. 93.

2. See G. F. Khil'mi, *Foundations of the Physics of the Biosphere* (Leningrad: Hydrometeorological Publishing House, 1967), p. 281; G. F. Khil'mi, in "Global'nye problemy," *Voprosy filosofii,* 1974, no. 9, p. 82; and M. M. Kamshilov, "Chelovek i zhivaia priroda," *Priroda,* 1969, no. 3, p. 33.

3. For more on these debates see my piece, "Optimists and Pessimists: The Ecology Debate in the USSR," *Canadian Slavonic Papers,* vol. 26 (1984), no. 2, pp. 127–40.

4. This notion is explicitly stated by the East German economists Horst Paucke and Günter Streibel in "Zur Verflectung von Naturprozessen und volks-wirtschaftlichen Reproduktionsprozess," *Wirtschaftswissenschaft,* vol. 28 (April 1980), pp. 404–409.

5. See, e.g., V. G. Markhov, "Naucho-tekhnicheskaia revoliutsiia i prirodnaia sreda," *Voprosy filosofii,* 1974, no. 8, p. 100; and A. A. Arakelian, "The Scientific-Technical Revolution and the Biosphere," *Voprosy ekonomiki,* 1976, no. 5, translated in *Problems of Economics,* vol. 19 (March 1977): pp. 73–77.

6. *Okhrana okruzhaiushchei sredy i ratsional'noe ispol'zovanie prirodnykh resursov v SSSR: statisticheskii sbornik* (Moscow: Gosudarstvennyi komitet SSSR po statistike, Informatsionno-izdatel'skii tsentr, 1989), pp. 76, 108.

7. For further discussion of these policies and some early results, see De-Bardeleben, *The Environment and Marxism-Leninism,* pp. 59, 162–63.

8. The discussion that follows draws heavily from my book, *The Environment and Marxism-Leninism,* pp. 189–90.

9. Geoffrey Hosking's book, *Awakening of the Soviet Union* (Cambridge: Harvard University Press, 1990) suggests a tradition of such methods of local adaptation in other dimensions of Russian life. See especially chapters 2–3.

10. See, e. g., criticism of proposed petrochemical complexes in Tiumen' by Valentin Rasputin, "Giants: Hard on the People, the Pocketbook, and the Environment," *Moscow News,* 16–23 April 1989, p. 15; and the discussion in *Kommunist,* 1989, no. 1, pp. 23–33, and no. 5, pp. 76–77.

11. *Pravda,* 24 December 1990, p. 3 (translated in *Current Digest of the Soviet Press* [*CDSP*], vol. 42, no. 51 [1990], p. 18).

12. See Kaunas Economic Institute, "Urgent Ecological Problems in Lithuania," A Brief Submitted to the Council of Ministers of the Lithuanian Soviet Socialist Republic, November 1988, pp. 11–15.

13. See. e.g., "Brief of the Lithuanian World Community and the Lithuanian Green Movement to the Conference on Security and Cooperation in Europe: Meeting on the Protection of the Environment," Sofia, Bulgaria, October 1989.

14. On a successful referendum to close an agrochemical association in Odessa see *Izvestiia*, 23 December 1990, p. 6; on a referendum in Cheliabinsk to block the South Urals Atomic Power Station and to stop burial of radioactive wastes in the Southern Urals, see *Izvestiia*, 8 December 1990, p. 2.

15. See Tatiana Zaslavskaia's interesting analysis in *Izvestiia*, 17 December 1990. Referring to the local level, she suggests that "it appears that the democrats were too hasty in taking power into their own hands. . . . political power must invariably be backed by economic power" (translated in *CDSP*, vol. 42, no. 50 [1990], p. 10).

16. See the contribution of Susan Linz in this volume for discussion of this topic.

17. On this subject, see my article, "Environmental Protection and Economic Reform in the USSR," *Soviet Geography*, vol. 31, no. 4 (April 1990): pp. 237–56.

18. For an excellent discussion of these and other issues related to specific types of technology see Philip K. Pryde's book, *Environmental Management in the Soviet Union* (Cambridge, England: Cambridge University Press, 1991).

19. Gosudarstvennyi komitet SSSR po okhrane prirody (hereafter Goskompriroda), *Doklad: sostoianie prirodnoi sredy v SSSR v 1988 godu* (Moscow, 1989), pp. 82–83.

20. Ibid., p. 158; on Uzbekistan, see N. Skripnikov, "Khimizatsiia, zdorov'e i zakon," *Pravda vostoka*, 4 March 1989; on Tadzhikistan, see "O 'bezopasnykh pestitsidakh'," *Nedelia*, 1988, no. 47, p. 3; see also "Pestitsidy: vred i pol'za?" *Argumenty i fakty*, 1989, no. 11, p. 6.

21. "Brief of the Lithuanian World Community," pp. 12–13.

22. Goskompriroda, p. 83.

23. Ibid., p. 88. See also *Okhrana okruzhaiushchei sredy*, p. 101.

24. Goskompriroda, p. 188.

25. *Okhrana okruzhaiushchei sredy*, p. 100.

26. See, e.g., "V postoiannykh komissiiakh Verkhovnogo soveta TSSR," *Turkmenskaia iskra*, 6 May 1989, p. 3.

27. *Izvestiia*, 18 December 1990 (translated in *CDSP*, vol. 42, no. 51 [1990], p. 13).

28. *Pravda*, 24 December 1990, p. 3 (translated in *CDSP*, vol. 42, no. 51 [1990], p. 18).

29. Goskompriroda, pp. 93–94, 187, 191.

30. Goskompriroda, pp. 62–74, 94–99.

31. *Okhrana okruzhaiushchei sredy*, p. 156.

32. Ibid., p. 153.

33. Ibid., p. 154.

34. Goskompriroda, p. 77.

Technology and the Environment in Eastern Europe

Barbara Jancar-Webster

Technology is a Pandora's box and therein lies its paradox. It is responsible for dirtying our environment but also responsible for cleaning it up. When we look at the environmental crisis in Eastern Europe, we are inclined to lay the blame rather too easily on the political system. Over the past year, the mass media as well as the scholarly press have been actively publicizing the extent of the crisis, as rapidly as the data come in. The East European countries have also published official studies of the state of their environment. Given the availability of information, no purpose will be served by reviewing the figures; the reader is referred to the government documents listed in the bibliography at the end of the paper for the most recent government statistics from Hungary, Czechoslovakia, Croatia, and Slovenia (see also Matas, Simoncic, and Sobot 1989).

The East European systems are not the only ones in which severe environmental pollution has occurred. The West as well as the East has its share of problems. This paper proposes to explore the general relationship between technology and the environment and then to apply the findings to the current situation in Eastern Europe. The paper hopes to show that the transition to democracy through which the area is now passing provides both advantages and disadvantages in terms of improving environmental conditions. If foresight is shown in choosing development technologies, hopefully the advantages will prevail over the disadvantages.

Technology and the Environment:
A Hypothetical Framework

In a consideration of the relationship between technology and the environment, four central features stand out. The first is that throughout history the type of technology prevailing at any time has been closely associated with the political, social, and economic system. Like all systems, the social system is an ever-changing mix of stability and resilience, conservation and adaptation. In the past, systemic tendencies to stability have prevailed, unless surprise in the form of violent overthrow or technological innovation has intervened. Today, the pace of change has quickened. The microchip revolution has facilitated not only the accurate monitoring of our economy but the precise documentation of the impact of our technology on the environment. The problem is that the system that produces our pollution has shown itself to be particularly intractable and inflexible in its responses. Davis (1990, p. 57) argues that system rigidity has its origins in the energy infrastructure for a number of reasons: power plants may last as long as forty years; long lead times are necessary to develop new energy projects; and there are entrenched public perceptions of cost, need, and environmental acceptability. When we talk about changing an energy technology, we are talking about changing an entire economic and social system.

The second fact is that civilization as we know it has evolved from the exploitation of the most readily available and least costly (in terms of energy needed to access it) energy use linked to specific technological inventions, to increasingly costly energy systems linked to new technological inventions. All agricultural systems were based on the use of wood. Civilizations such as Mohenjo Daro and Harappa went out of existence when the fuel source was gone and erosion had taken its toll. In Europe in the Middle Ages, most daily products were made of wood, from mixing bowls and butter churns to mine shafts. England was depleted of most of its virgin stands before the Christian era. By the end of the twelfth century, to keep their profitable mines open, the Czech kings were forced to reforest denuded woodland by importing the fast-growing Alpine fir. These now dominate the Czech landscape. Demand for wood was a driving force behind the English encouragement of colonies in the New World, where woodcutting proceeded at an unprecedented pace. By the Battle of Yorktown, huge sections of the once heavily wooded eastern seaboard had become open fields. In the eighteenth century, wood had become a scarce resource, and England turned to coal.

Coal required much higher concentration of resources. The key invention, the steam engine, made mass production economic by drastically reducing the cost of land and sea transportation. Hodge (1990) suggests that the Romans had the technological means to develop mass production, but did not do so because they were not able to solve the problem of high transportation costs. The steam engine pumped the water out of the mines, transported the coal to smelters, and powered the iron smelters. Smelters provided the iron for construction of the steam engine, the rails on which the engines and wagons traveled, the mining equipment, and ultimately the ships that rapidly disseminated the new products of the industrial revolution.

If the concentration of resources at the mines and in the new factories, necessitated by the powering of the steam engine with coal, made rapid industrialization possible, it also contributed to the rise of the modern polluted and polluting city. Philosophers and poets such as Thoreau, Wordsworth, and Marx deplored the "dark Satanic mills" and "the getting and spending" that accompanied the rise of capitalism.

The second phase of the industrial revolution began at the end of the nineteenth century with the developments of electric power, petroleum, and natural gas as energy sources. The technological inventions that accommodated these new energy types were the internal-combustion engine and the chemical and metallurgical industries. The age of the automobile and airplane saw a network of roadways leading to and from airports, and around and into the cities, crisscrossing the land. To escape the pollution of the inner city, the better off moved out, first along the railroad lines and then along the newly built roads. The suburbs came into being. In the United States, where cities tended to develop as economic rather than cultural centers, the flight of the middle and upper class brought urban blight and inner-city ghettos (Watson, and O'Riordan 1976, pp. 79–92). The development of plastics generated the throwaway society. The new kinds of energy required even greater outlays of capital than that needed for coal, to build the pipelines and power lines to transport fuel to the new sprawling megapolis. Slowly but surely the spaces between the cities filled in with the growing population, turning the East and West coasts of the United States, as well as most of Western Europe, into one vast suburbia.

The microchip, biotechnology, and advanced materials ushered in the third phase of the industrial revolution. In a few decades the computer has revolutionized society. The information and automation age has globalized national economies, enabling us to transmit data around the world in seconds, linking national stock markets, bringing parts made in one region of the globe for assemblage in another and sale in yet another. Equally important, we have gone to the moon and seen "spaceship earth" in its entirety from afar. No longer is the earth the background, nor is man the lord of all he surveys. Instead, the universe has become the background, and the biosphere is now the dominant figure (Timmerman 1986). Technology has forged a global interdependence that has undermined our once stable concepts of national sovereignty and international behavior (Caldwell 1973; Mische 1989; Smith and Vodden 1989) and heightened our vulnerability as passengers aboard a fragile and unique planet.

The third fact of the relationship between technology and the environment is that, as Brooks (1986, p. 329) describes, the benefits of technology often increase in proportion to its scale of application, whereas environmental and social disjunctures resulting from the application of technology increase nonlinearly as the scale of application increases. By the time the negative environmental surprise registers in public consciousness, society is already accustomed to the technological benefit. Brooks argues that as technology matures it tends to become more homogeneous and less innovative and adaptive. Success freezes the corporate enterprise into a mold dictated by fear lest departure from a successful formula jeopardize aggregate capital investment, marketing structure, and supporting bureaucracies. In the early stages of a technology, there are many options and choices. Competition dominates the market. Gradually one technological variation begins to win. Economies of scale, marketing, and production give it a competitive edge. Technical options become narrower, and research tends to be directed only at marginal improvements. The technology has produced what Brooks terms a *technological monoculture*. The new technology and its supporting systems (the marriage of the automotive and petroleum industries, or the petrochemical industry) now constitute a more and more self-contained social system, unable to adapt to the changes necessitated by its success (Brooks 1973, p. 253). Brooks concludes: "Thus, the technological activity may become strongly established with influential vested interests during the linear regime, before the disbenefits that increase non-linearly with scale become . . . apparent to the wider public" (Brooks 1986, p. 329).

In a 1976 study of modern industrial society, Slovak economist Eugen Loebl (1976, pp. 140–41) came to a similar conclusion. In his view, what started as a competitive market situation has become highly uncompetitive. The actors in the spheres of production have become much stronger than the consumers. The whole economy is now geared toward increasing the amount of goods and services to be consumed. The consumer has lost any ability to influence the market and, instead, has become an object of manipulation by the producers. The resulting consumption-driven society organized in the interests of the producers has proved incapable of dealing with problems of pollution. In his words, "the problems of air and water pollution and of the growing scarcity of resources ... cannot be solved by thinking of input and output only in terms of price units."

A final factor is that all human action on the environment tends to reduce variability. The agricultural revolution focused on raising some grains to the exclusion of all others. The domestication of animals was a selection of the most tractable of the wild creatures. Technology is a selective process. By contrast, natural ecosystems, among which human society may be included, are marked by high variability. Scientists are able to classify and describe these systems by identifying what they term the slow, intermediate, and fast variables within them (Hollings 1986, pp. 301–7). For example, a savanna may be described in terms of grasses (fast variable), shrubs (intermediate variable), and herbivores (slow). Human concentration on any one of these defining variables leads to distortions in the functioning of the other two. When semiarid savannas were turned into cattle-grazing fields in the Sahel (focus on the slow variable), changes in grass composition occurred with an irreversible switch to woody vegetation (intermediate variable). Distortion in an ecosystem may influence changes in the global biogeochemical cycle, Lovelock's (1979) "gaia" effect. In this case, overgrazing and a shift in vegetation may reduce moisture in the air. Drought, in its turn, may trigger the collapse of such a system and the phenomenon of desertification.

We thus cannot see nature as neutral, or simply acted upon. Human management, in its effort to maintain the steady state of a natural ecosystem, provokes a response from the ecosystem as it tries to adapt its patterns of behavior in the face of disturbance. The Yorkshire countryside in England has been so long grazed that all the roots of the original forest have died. Given the rugged climatic conditions of the British Isles, it is difficult for trees to take root and live. The moors are nature's adaptation to the impact of human society. Because of the delay between our selective management (stability) and nature's response (resilience), we become aware of nature's reaction through the element of surprise. Hollings (1986, pp. 310–13) suggests three kinds of surprise: unexpected discrete events, such as Chernobyl or Love Canal; discontinuities in long-term trends, such as industrialization, technological rigidity, environmental pollution; or the sudden emergence of new information into the political consciousness, such as global warming, spaceship earth, or acid rain. Discontinuities may not manifest themselves in a single generation. One of the problems with the attempt to predict the future is the knowledge that the negative by-products of our rough predictions may not be known until they appear as "surprise" to future generations (Orr and Soroos 1979, pp. 327–43).

Our ability to respond to these surprises depends on the resiliency of our human institutions. The tendency for technical monoculture to develop huge support bureaucracies, which minimize risk and enhance survival (Venerable 1987, pp. 81–110), increases the stability of the system at the expense of resilience. On the one hand, the market is incapable of handling resilience under conditions of monoculture. Some nonmarket force seems necessary. Proponents of California's Initiative 128, "Big Green," argue that the only possibility for societal adaptation to altered environmental conditions is to change the game plan by new draconian regulations, such as those Big Green proposes. Technology is then free to be once again innovative (Ted Smith, talk at *EnSol90* 1990). Opponents say the changes are too disruptive and would radically change society, resulting in a society that is unable to adjust to the shock. What is lacking from all societies is a built-in promoter and guarantor of technological diversity. In many ways, technological diversity may be compared to genetic diversity. The existence of or knowledge about a multiplicity of technological options can be a source of systemic renewal and adjustment to surprise.

The problem is that, although some economists claim that environmental goods are capable of quantification (Pearce 1976, pp. 20–40; 1985, p. 16), public or common values remain outside the market-oriented efficiency criteria and organizational growth that drive the evolution of technological systems (El Serafy and Lutz 1989; Redclift 1987, pp. 36–51). The decrease in competition that accompanies technological maturation further erodes the prospects for technological diversity. In the third phase of the industrial revolution, the internationalization of the market and global interdependence may maintain competition where domestic oligopoly cannot, but this eventuality is by no means certain. Given the ever-increasing costs of research and development, domestic industries are not willing to assume the risk of new technologies unaided. What seems to be required is a profound change in public perceptions of need, cost, and life-style to secure government intervention on the side of technological experimentation.

To summarize, the interaction between technology and the environment occurs within human socioeconomic political systems or series of systems, which are ordered in time and space. The logic of technological development tends toward monoculture and conservatism after the initial period of innovation and exploration. Fortunately or unfortunately, the benefits of an innovation are experienced in direct proportion to its application, while the environmental disbenefits appear nonlinearly after the scale of application has reached a threshold. The disbenefits are perceived as surprises in the social and natural world. The increasing rigidity of institutions that occurs with the diffusion of the technological monoculture denies the industrial system the flexibility to respond quickly to surprise, whether it be discrete, discontinuous, or new information. Finally, human beings do not intervene in a neutral or static environment. Management strategies are selectively oriented, and depending on their orientation, the strategy will bring response from the other variables in an ecosystem, as the system seeks to adjust to the disruption of its intervariable rhythm. This distortion in turn can influence the biogeochemical global system, producing changes which then impact on human society and natural ecosystems. The figure on the following page, adapted from Hollings (1986), illustrates this process.

It is the assumption of this paper that the four features outlined above—the systemic and temporal factors, the linear/nonlinear relation of technological to environmental change, and the propensity of human management systems toward reduction in variability—characterize the general relationship between technology and the environment in all socioeconomic systems. However, differences in socioeconomic systems may contribute to changing the intensity or speed with which the environmental disbenefits manifest themselves by surprise and the population becomes aware of them. It may also shorten the time period within which the technological monoculture evolves toward paralysis and the system collapses.

A further assumption is that the relationship between the two can

Figure 1. **Interaction between Global Biogeochemical Cycles, Society, and Ecosystems**

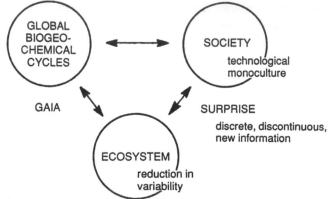

Source: Hollings, 1986, p. 309.

only be understood within the context of a global dynamic biosphere where the interruption of the rhythms of one system disrupts the rhythms of the other two, producing a cascade of interactions, rather like a bowling ball hitting bowling pins.

Technology and the Environment in Eastern Europe

In 1948, Eastern Europe fell victim to Stalinism. For Czechoslovakia, Stalinism meant deindustrialization. An advanced electro-engineering technological program was forcibly replaced by smokestack first-generation industrial production, coal, iron, and steel. In Poland and Hungary, a centrally planned economy and monopoly political system were the instruments of increasing the industrial tempo and collapsing the time between the first and second phases of the industrial revolution. In Romania and Bulgaria, Stalinism effectively launched industrialization. The decision in favor of economic autarky in the fifties meant that each country had to produce its own iron, steel, and aluminum. The key to Romanian growth was its refusal to follow a Comecon plan for agricultural development and its channeling of investment into metallurgy, machine building, energy, and petrochemical production. Indeed, in its effort to earn greater independence from Moscow from 1961 on, Romania used its laxer environmental standards to attract foreign investment. By 1980, more than 55 percent of Romanian export trade was with the first or third world (Leonard 1988, pp. 148–53).

Eastern Europe as a whole is not yet at the threshold of the third phase of the industrial revolution. While the northern tier of countries may be at the brink, the southern tier is more likely moving to the end of the second phase. Because the third phase is not yet stabilized anywhere in the world in a permanent institutional form, the introduction of the new technology can only be achieved by innovation and competition. The technological monoculture inherited from the Stalinist system proved incapable of providing the necessary climate for innovation, and the system collapsed. The entire area is now in a phase of economic and political adaptation to social surprise.

The system also collapsed under the weight of environmental surprise. In the Communist system, information about the environment was considered a matter of national security and kept secret. The West was subjected to well-orchestrated propaganda about how capitalism with its emphasis on the profit motive was responsible for world pollution. However, during the seventies, the nonlinear aspects of pollution in relation to technological development became increasingly visible. The first information to come to international attention came from the Soviet Union: the pollution of Lake Baikal and the Siberian rivers project. Demographer M. Bednyi suggested that mortality rates in Soviet cities might be influenced by the increase in environmentally related diseases (Bednyi 1984; Jancar 1987, pp. 289–94).

While data were appearing in the scientific literature on environmental pollution, most of it was highly restricted in its dissemination, and the public heard little about it. For example, under the U.S.-Polish cultural-exchange agreement signed in 1958, Polish scientists in the Institute of Occupational Medicine in the Textile and Chemical Industries and the Krakov Medical Academy, among others, received funding from the National Institutes of Health (NIH) for studies on chest diseases in coal miners, the effects of carbon disulfide on miners, the cumulative effects of long-term inhalation of lead on the respiratory system, the health of workers in the iron and steel industry, and the effects of industrial poisons and drugs on worker health (Jancar 1990a). Reports on all these studies were filed with NIH and the cooperating Polish research institute, where inertia virtually buried them. In 1980, an important scientific conference was held in Prague that gave a frank evaluation of the worsening environmental situation (Riha 1980). Vintrova, Klacek, and Kupka (1980) published a subsequent article in the economic journal, *Politicka ekonomie,* baldly stating that unless environmental protection were integrated into the economic plan, the country would reach the ecological barrier where further economic development was impossible. The conference report was classified for distribution to top government policy makers. The professional journal circulated only among specialists. In Hungary, a special scientific committee recommended against the Nagymoros Dam in the mid-eighties, but the members were apparently so intimidated by the government that the report was silenced, and scientists are only now saying publicly that they had participated in the committee findings.

Mass dissemination of environmental information started in Poland with the advent of Solidarity. The disaster at Chernobyl increased public anxiety and awareness of environmental problems, but the full force of information in Czechoslovakia, Bulgaria, East Germany, and Romania did not reach the general population until 1989. Now the world knows that the damage done to the environment by technological monoculture under Communist systems may be even worse than in the West. It took only thirty years for the discontinuity between technological application and environmental response to manifest itself, mainly because, rhetoric notwithstanding, few effective steps were taken to control it (Jancar 1989, 1990b).

In her study of environmental management in the Soviet Union and Yugoslavia, Jancar (1987) analyzed the structural reasons for the catastrophe. Foremost among these should be mentioned the regime's absolute monopoly over the state, law, economy, and information. The political support of technological monoculture, of innovation stifled by monopoly, lengthened the lead time from blueprint to application of new technology and fostered a pattern of bureaucratic infighting between state-owned polluters and state-appointed environmental enforcers, which could continue unabated in the presence of a largely ignorant and depoliticized population. Exhibit No. 1 of the nonlinear impact of bureaucratic incompetence and human error was Chernobyl. In effect, the social and environmental disadvantages accruing to the successful application of a technology in a free society were both intensified and hastened by a prior decision to freeze the whole system into a mold informed not by market success, but by ideological principles. Because the economic system provided the security and control function that maintained the ruling group in power, the Communist leaders found themselves in a situation where they could not innovate in the system without bringing the whole system down. While economic restructuring was necessary to be competitive on the world market and to improve living standards at home, regime stability depended on the maintenance of monopoly. By forced reduction in variability and insistence on homogeneity, the Communist system interrupted the normal rhythm of technological development and destroyed civil society. Administrative paralysis set in, multiplying the negative impact of ecosystem surprise. Nowhere in the world is the ecological barrier as proximate a reality as in the former Communist countries.

Eastern Europe is now in the stage of what Hollings calls "creative destruction." (Hollings 1986, pp. 307, 312). The old society is dead, but the new has yet to take form. Concern for the environment is on the lips of every politician. But the fear is that popular expectations of improved living conditions supported by the governments' urge to become full players on the world market will result in the area's swift integration in the prevailing global technological monoculture at the expense of environmental remediation.

The Disadvantages

The transition to a new society presents Eastern Europe with advantages and disadvantages in its effort to adapt technology to a ravaged environment. To take the disadvantages first, foremost among these is the fact that East Europeans see themselves as part of European culture. The end of the twentieth century has witnessed a homogenization of expectations to Western standards among all peoples of the world. One of East Europe's greatest dreams is to be fully reintegrated into its parent culture. Reintegration means adopting the affluent life-style that the former Communist regimes called decadent: More money equals more cars, more appliances, more consumption, more convenient energy, more electricity (Fri 1990, p. 8). To argue that one does not need all these things, or that consumption is destroying the Western environment, is nonproductive. East European societies have been too long subjected to scarcity and deprivation. Renewal must be accompanied by an improved standard of living along Western European lines.

A second disadvantage is that the adoption of new environmentally friendly technology is not just a question of being able to buy and install. The application of new technology affects the entire infrastructure, including education, training, communications, and transportation. While East European science has excelled in several areas, re-search funds have been traditionally limited, and excellence has been concentrated in selected research institutes. Moreover, the old system was very inefficient in applying new scientific developments, whether developed at home or purchased from abroad. A *Wall Street Journal* analysis (15 January 1990) cited the continuing rigidity and unrespon-siveness to rapid change of Hungarian enterprise organization. Many CEOs do not have a clear understanding of strategic planning, and managers tend to be technological followers. Nowhere in Eastern Eu-rope is there the managerial expertise in environmental analysis. To change old habits and *modi operandi* will take time and the willingness and opportunity to learn.

Equally important, the development of most environmentally friendly technology is costly, the benefits as yet unclear, and commer-cial availability still in the future. Pollution control is a thriving state-of-the-art industrial sector. Since the industry is in its pioneer phase, a great deal of investment in research is required, many products are experimental, and prices are correspondingly high. Torrens (1990, pp. 29–30) estimates that for a new coal-fired power plant emission-control costs typically amount to about 30 percent of overall plant capital costs and can reach 40 percent if the plants have to meet strict standards. Most of the clean-coal technology is still in the demonstra-tion phase, and availability for commercial orders is not expected until 1991 and later. The East European countries are in no financial posi-tion to be able to purchase these technologies, and governments may be reluctant to do so in the face of uncertain benefits. There is a very real danger that a clean environment will become a luxury for the wealthy countries, which the poor countries of Eastern Europe and elsewhere cannot afford.

A third disadvantage is that the old bureaucracy still remains in power. Under the Communist regime, as an extension of the state, the economy shared in the political-patronage game, promoting the loyal, firing the unreliable, and permitting the emergence of no enterprises or organizations that could acquire sufficient independent economic means to challenge the ruling hierarchy. Virtually every individual in a position of administrative or managerial authority in Eastern Europe and the Soviet Union today was at one time either a member of the Party and/or a client of a powerful patron in the ruling group. Although the infamous *nomenklatura* system has been formally abolished in

Poland, Hungary, and Czechoslovakia, most of its members still occupy almost all the important positions. Oligarchic and patron/client networks may have been somewhat broken up, but they are still functioning. Hankiss (1989, p. 8; 1990, pp. 203–206) sees the transformation of the former Communist "vanguard" into a new *grande bourgeoisie*, as Party and state bureaucrats exchange their positions in the old bureaucratic hierarchy and move into the managerial and entrepreneurial spheres. During the transition period of "creative destruction," bureaucratic power has become convertible and is being converted into economic positions and assets that, in addition to providing more income, will assure that power and status is transmitted to the next generation. The closest Western analogy to this emerging coalition is the French administrative elite, with its family connections and domination of the *grandes écoles*. System renewal risks the perpetuation of the old rigidities under new labels. The new East European bourgeoisie may shed its bureaucratic intransigence for an analogous egocentric concern for economic survival and choose to opt for the "safe" technologies that have conquered the globe, rather than encourage innovation and experimentation.

The fourth disadvantage is the existence of contradictory goals and values. There is ambivalence about the market economy and little understanding of democracy. Consumerism and private property so far have not been capable of motivating a vibrant economic renewal. Poland's "plunge" into capitalism has proved to be icy indeed, and the discussions drag on in Czechoslovakia and Hungary as to the speed of the transition to a full market economy. In Yugoslavia, the collapse of the Communist monopoly has brought the old national rivalries tragically to the surface, propelling the country toward disintegration rather than economic revival. When one considers the fact that the new values are being propagated by the old bureaucrats and former Party members with the same enthusiasm as they propagated the Marxist collectivist values, it is no wonder that the people are cynical and to a large degree politically unmotivated. In Czechoslovakia's June 1990 elections, people were told to cast their vote for Civic Forum as the vote against communism. However, on the eve of the election, the new party behaved no differently than the unpopular Communist party in using smear tactics to turn voters away from voting for a leading political rival. While older people with a memory of the First Republic wept when they went to the polls, young people observed by the author voted without any show of conviction or enthusiasm, as if they had little understanding of what the right to vote meant. Only 30 percent of the population participated in Hungary's local elections on September 30, 1990. The tenth in the space of a year, all they could offer was a continuation of crisis and pain. The election results suggested that voters cared more about nationalism than about democracy.

Finally, the destruction of the old regime may result in a new colonialism. The share of pollution-abatement costs in obtaining raw materials or producing a product has been steadily rising in Western Europe, Japan, and the United States. Leonard (1988) has documented some transfer of polluting industries by the multinationals to lesser developed countries. Representatives of the U.S. mining industry claim that the transfer has already occurred in their business. Increasingly restrictive environmental regulation has so raised the cost of mining in the United States that there are only five hundred mining sites left in operation (*EnSol90* 1990). East Europeans also fear that the multinationals may see Eastern Europe as a pollution haven. Desperate to raise standards of living, the new East European governments may accept lax enforcement or lower standards to ensure employment and economic renewal. The author's discussions with national officials were not encouraging. Most insisted that it was impossible to renovate the economy under conditions of strict environmental regulation. There had to be a trade-off. The first priority was to get the economy going. After that, one could talk of environmental protection.

The Advantages

These disadvantages are offset by several advantages. In the first place, the old system is seen as the cause of both environmental and economic failure, with the former clearly linked to the latter (Komarek 1988, pp. 306–10). The public's perception was summed up in the environmental movement's catchwords—gigantomania, megalomania. In Czechoslovakia, Hungary, and Yugoslavia, the building of dams exemplified the Communist regimes' arbitrary methods of rule. In Hungary, the Nagymoros Dam became the focus of the fight to bring down the old system, while in Yugoslavia, the defeat of the construction of the dam across the Tara River Canyon marked the coming together for the first time of environmental interests crossing republican lines. Significantly, the environmentalists used the same vocabulary as did John Muir in his long opposition to the Hetch Hetchy Dam at the beginning of the century (Chowder 1990). In Slovenia and Croatia, gigantomania was seen in the construction of the Krsko nuclear power station. The Slovenian minister of energy, I. Tomsic, told the author that Krsko represented the official attitude toward industrialization at the time: big projects, big energy, big industry. The same wording was used by an employee at the newly established Czech Ministry of the Environment (MZP) in reference to the Czech nuclear-power complex under construction at Temelin in South Bohemia.

The first chapter of the Czech Ministry of the Environment's report on the status of the environment in the Czech lands (Moldan 1990, pp. 15–18) identifies the three principles of *totalita* that negatively impacted upon the environment: The first was the system's continuing efforts to strengthen its power, the second was its "system of collective irresponsibility," and the third was the system of branch ministries where every human activity was administered by ministerial bureaucrats with unlimited power. In effect, the old regime symbolized the most negative aspects of technological monoculture. The pursuit of big, environmentally unsound economic projects was perceived as the logical outcome of *totalita,* where bureaucrats make the decisions and the public has nothing to say. The East European public would agree that it is the nature of the single-party monopoly system to push economic programs destructive of the environment.

A second advantage of the transition period is that it represents the interim between one system and another. There is a chance to change things. The East European countries do not have to imitate the West's industrial pattern. They do not have to get locked into the West's political/economic quarrels. They can decide to integrate environmental and economic-development goals or to promote consumer-driven democracy, rather than what Loebl (1976) termed the West's consumption-driven society. The collapse of one system does not mean that only one alternative is available. There are other untried options. Renewal means experimentation. After the Second World War, Japan was able to leapfrog into the forefront of industrial development. Leapfrogging is certainly a possibility for Eastern Europe, since all the countries are now free to determine their own future.

The one factor that might blind the region to other options is what Galbraith (1990) terms simplistic ideology. If communism was bad, capitalism is good. If communism falls, capitalism is triumphant. The

Marxist utopia collapses before the capitalist heaven. The future is yours, Mr. Ivan or Miklos Carnegie, if you can weather the storm until the transition to capitalism is complete. Galbraith and Loebl both argue that modern Western society bears little comparison to nineteenth-century capitalism and none at all to Western economic theory. Privatization provides no panacea for Eastern Europe and, as Western society illustrates, no buffer against the increasing and overriding role of organization and bureaucracy. For power in Western society has been transferred from the private owner to corporate management. What is argued here is that under socialism, the bureaucratic underpinnings of technological monoculture were too rigid, too inflexible to survive the combined assault of a Western technological organization driven by success and environmental surprise. However, it is by no means clear that Western society is sufficiently more flexible to adapt to environmental surprise. Far from vindicating a free-market system, the collapse of Communist planned economies focuses attention on the existence of parallel vulnerabilities in the West. Eastern Europe now has the unique chance to study the mix of rigidity and resilience in Western society, before it chooses its technological imports or experiments in the development of its own technologies.

A third advantage is the state of East European industrial production. In most cases, it will be better to replace rather than repair outmoded plants. Replacement opens up the choices of plant reformulation, redesign of equipment, and process modification. Replacement offers the opportunity to build new plants with pollution controls in place or the integration of pollution controls in new plant design. Experience has shown that adding scrubbers to a coal-fired plant can cost up to 45 percent of capital investment. Integrating controls at the design stage can reduce costs by almost half. Studies (Cleaning Up . . . 1990; *EnSol90* 1990, Plenary Session, pp. 1–7) suggest industry tends to find pollution prevention less costly than end-of-pipe control. East European industry has the opportunity to design from the ground up. However, end-of-pipe technologies are more available and more sure and, thus, are preferred where environmental legislation mandates the best available technology. In addition, Eastern Europe is full of domestically produced items that no one wants to buy, and the region is being flooded with products from the West. Here could be another unprecedented opportunity to design and produce fast-degradable products or products with minimal environmental impact at end of life. A new beginning means the industrial mold is not yet set. Innovation can play a role in selecting strategies for environmental management. For example, rather than mandate an emission standard based on BAT (the best available technology), legislation could put a graduated tax on excess emissions to encourage industry to look for a suitable technology at the lowest price.

There are signs that East European governments are already considering different options. One of the most widely discussed options is providing economic incentives for environmentally safe technology. Among these are the pollution tax, the environmental fund, the requirement of an environmental-impact statement in economic-development design, the linking of loans and other means of finance to environmental requirements, restructuring the price system to reflect the cost of environmental inputs, and valuating environmental damage (Czech Republic 1990, pp. 145–47; Republic of Croatia 1990, pp. 96–103).

Last but not least, there is the well-developed connection between environmental change and democracy. This connection may be the biggest advantage the East European countries have going for them. The thirst for democracy is in all aspects of life. People are not sure what democracy means, but it would seem to cover a manner of behavior that goes beyond voting for a representative at election time.

Under the old regimes, environmental groups were in the forefront of the movement for democracy. In their search to remain within the law, yet bring public pressure to bear on environmental problems, the environmental groups pioneered democratic methods of political action. Foremost among these methods were Samizdat publications (Charter 77), letters signed by thousands of petitioners, the simultaneous demonstration in several towns at once, the massed street parade, securing the support of international environmental organizations, and cooperating with them in international actions.

In Poland, the Polish Ecology Club was the first group to form in Kracow. Its initial action was highly symbolic. With the backing of an understanding mayor, the group was able to force closed the Skawina aluminum plant, one of the major industrial polluters of the area. For years, the Hungarian Blues walked a tightrope between legality and illegality. According to Judit Varsanhelyi of the East Central European Environmental Center in Budapest, the Danube Circle was the first organization in Hungary to hand out handbills in the street. When the group placed advertisements in foreign newspapers, the government accused them of going back to "the ugly thirties" and using foreign support to influence national politics. But the movement persevered. In the spring of 1988, the group organized one demonstration at Nagymoros itself. It was supposed to be a march of women and children. But so concerned were the fathers about the safety of their families that they went along. The women handed petitions through the construction gates and chatted with the engineers and workmen. The demonstration turned into a celebration of family life.

In Yugoslavia, the antinuclear movement, started by a young student in a Belgrade high school, expanded across the country with groups organizing in every republic. Cooperation rapidly spread between groups across republics. The movement became so powerful that what the press termed the pronuclear group was forced to organize public debates to give their side of the problem. Success was reached when the Yugoslav Youth Organization introduced a resolution in the federal Skupstina requesting a moratorium on all nuclear power until the year 2000. The Skupstina vote in favor of the moratorium was the first time the public had directly influenced a legislative decision at the federal level. In Slovenia, the ecological movement was started in the early eighties by a small group of students who one day decided to exercise the rights given them by the Slovenian constitution. According to Peter Jamnicar, secretary general of the Greens of Slovenia, the regime was caught off guard. Never before had citizens protested so directly in public. In 1989, the movement decided to form a political party, even though there was no law legalizing political parties. The law came in January 1990, and the Greens went on to win 8.8 percent of the popular vote in the first free elections in Slovenia. The vote gave them representation in Parliament, four ministries, and one seat in the Slovenian presidency.

In Czechoslovakia, Romania, and Bulgaria, environmentalists exercised their legal rights much more hesitantly because persecution, violence, and imprisonment remained regime responses up to the very end. Nevertheless, despite active measures by the authorities to repress demonstrations, in 1988 citizens in north Bohemia decided to send a petition to the local authorities protesting pollution in the area. It was the first spontaneous action in the region. Another two hundred braved police brutality in November 1989 to protest the appalling levels of air pollution in Khomutov on the Czech-German border, just a week prior to the November 17 demonstration in Prague that brought down the regime. In Prague, a group of women calling themselves the "Prague Mothers" risked imprisonment to protest the unclean drinking water they could not give their children. In Bratislava, scientists went public with a report (Budaj 1987) on the status of the environment in the Slovak capital. The report linked the rising suicide rate with the high levels of pollution. The first popular demonstration in Bulgaria took place in the Danubian city of Ruse. At issue was the lethal air pollution coming across the river from a chemical plant in neighboring Romania. In every country, the ecological movement developed and taught democratic methods of protest to large numbers of people.

Many tactics were taken over from West European environmental groups, especially the Greens, but the choice of methods and the decision to use them were made at great risk locally. Even in more liberal Hungary, activists who took to the streets faced loss of permission to travel abroad, denial of visas to children, and daily harassment from the authorities. Organizing protest measures under conditions where protest behavior was considered treasonable was dangerous and exhausting. In no country could activists make broad appeals in the newspapers or mass media. Signatures had to be gotten carefully, with the petition circulating from trusted person to trusted person. The coming of glasnost did not ease the organizational burden. The school of democracy was a full-time occupation. To quote Judit Varsanhelnyi once more, "our children were made orphans by the movement" (Jancar 1990c).

Despite the enormous dangers and the physical exhaustion of the activists, the environmental movement grew in each East European country and was one of the main catalysts in bringing down the regime. The message was simple and direct. *Totalita* had degraded the environment. Only democracy could restore it.

With such a record, the environmental movement is in a good position to influence the course of democracy and technology in all the East European countries. However, persecution and harassment notwithstanding, it was relatively easy to motivate diverse groups to protest against gigantomania. The environmental sins of *totalita* were clear and obvious. It will be harder to promote a positive program that will not appear to the general population to be a recipe for continued deprivation and poverty. In terms of our model, immediate investment in economic infrastructure is the "fast" variable. Jobs and life-style are the "intermediate" variables. Economic growth is the "slow" variable, whose social and environmental surprises are only understood in the far future. Restoring the damaged environment is a long process. It will be difficult to convince an untested government that the choice of environmentally friendly technology will pay off in the long run, particularly when the up-front capital costs are high.

The link between the environment and democracy goes deeper than the value of pressure groups to a democratic society. On the one hand, studies suggest that an environmental perspective may change the way industrial management behaves. There is evidence (Venerable 1987, pp. 77–80; Cleaning Up . . . 1990) that an environmentally concerned management communicates more with its employees, solicits their innovations, and cares more about the quality of the product. Environmental concerns foster new types of business contacts. They can establish new links between a company and its customers, with the focus shifting from sales to service. One such link might be the company's agreeing to dispose of the waste generated by the consumer's use of the product. Equally important, they motivate enterprises to develop community relations (Westerbeck, in *EnSol90* 1990, pp. 459–72). And environmental concerns can forge new connections among companies to pool resources for recycling or some other environmental project.

Legislation requiring the manager to collect and report pollution data encourages management's development of an environmental perspective. For many firms, the required data are the first objective information that management has ever seen on how much the company pollutes. An environmental audit then becomes of invaluable assistance in determining the cost of pollution remediation and incorporating the cost into prices. Under the Communist system, the official environmental organizations in Eastern Europe cooperated in collecting environmental data and will continue to do so. By cooperating with the environmental groups in the data collection and analysis process, industry can become more sensitive to the impact of pollution on product cost and on society at large. In making individuals more ready to protest perceived environmental injustices, the guarantee of civil liberties now in force also encourages enterprise management to consult with the community before it has chosen a technological course of action, rather than brave the public outcry in midapplication (Dickson, in *EnSol90* 1990, pp. 169–79). The promotion of an environmental perspective can thus bring democracy into the old, rigid hierarchical industrial organization. Innovative management behaviors can reduce production costs and make the product competitive in an international market where consumers are increasingly "green."

But no legislation or technology will be effective without the financial means to apply it. The new social systems are highly vulnerable to immediate surprises. The familiar Communist stability is gone. Resilience and adaptation are nonlinear, chaotic phenomena. The technological monoculture of the West seems safe and familiar. The path to Washington and the European capitals is already well worn. Conditioned as the people are to hierarchy and central leadership, the recent Hungarian and Polish elections indicate they are ill equipped to handle apparent chaos. In the face of uncertainty, public pressure may combine with the transformed bureaucracies to choose a prepackaged solution. The example of the West German takeover of East Germany is a case in point. In the race for a place in the global markets and an immediate higher living standard, East European societies may opt to forego the development of competition and innovation at home for immediate economic gains. With the world on the threshold of a new stage in the industrial revolution, which may weaken their technological hold over society, the multinationals are anxious to push their products on the vast, virtually untouched East European market. Once the trappings of a consumption society are in place in the East, it may be more difficult than it now is in the West to transform the life-style to do without them.

Elsewhere, I have urged that democracy and environmental restoration go hand in hand in Eastern Europe (Jancar 1990b). I would add a third factor—economic growth. The leadership can only be convinced that economic growth and environmental protection are linked through public pressure. Industrial management will follow the Western example of the three-month bottom line. Needing quick returns on investment, Japanese and Western industry may have little interest in environmental solutions. The destruction of the tropical rain forests from Malaysia to the Amazon are cases in point. The European Community, the IMF, environmental agencies of Western governments, and international NGOs can proffer suggestions, hold training seminars and scientific conferences, and distribute funding for environmental purposes. But in the East, as in the West, the public alone can compel governments to change the legislative game plan toward creativity in the import of environmentally safe technology and innovation at home.

Tied to the automotive monoculture, Washington is powerless to change fundamentally its transportation policy. European governments also protect the automotive industry because it provides jobs, earns foreign exchange, and provides public revenue. Governments have become as dependent upon the car as the Western consumer. To ensure future profits, the Western car manufacturer will design cars that will be required to be more environmentally friendly, but more important, will continue to reflect the owner's status and wealth. According to Boehmer-Christiansen (1990), status means large horsepower, interior comfort and space, and all the entertainment to help the driver while away his time in traffic jams. Increasing numbers of cars will eventually lead to restrictions on car ownership either through rising costs or regulation. The inevitable surprise comes when efforts to restrict car use set Western governments at odds with the major industries that undergird their economic survival. Eastern Europe is now at the crossroads. The choice would seem to be unequivocal. Be creative. Provide a transportation service not a product. Public transport now is cheap and available. It can be upgraded and modernized. But public transport will only be improved if the public demands it.

The environmental movement can play a major role in fostering environmental awareness and encouraging environmental solutions in a democratic manner. The efficiency with which it performs this function depends upon its ability to solve its own internal conflicts. The movement came to the forefront of the East European scene united in the focus on individual freedom and responsibility, but divided on the cause of pollution. According to Ivan Deimal, founder of the independent Czech environmental group the Ecological Society, the eighties was a period of considerable discussion among Czech scientists as to whether the environment was a *political* or a *social* problem. To state that it was a *political* problem meant challenging the regime. If the environment was considered a *social* problem, then in principle, the regime would be open to arguments and pressures for doing something about the deteriorating situation. In the words of Janos Vargha, "The environment recognizes no political system." This argument is not unlike the division in the United States between what Walter Truett Anderson of the Pacific News Service has termed "the politicos" and "the greens." The former consider the current system democratic enough and, hence, focus on getting laws passed and regulations implemented. The latter believe the environment can be saved only through a fundamental transformation of Western society from a corporate economic and political system into a more just and ecologically safe community through the practice of what Barry Commoner calls ecological democracy (Shades of Green . . . 1990).

The scientists who have come over from the academic institutes into the newly formed ministries of the environment essentially represent the view of the "politicos." Now that democracy is in place, environmental protection is mainly a matter of problem solving: passing the right laws and seeing that they are implemented through the best available technology, the provision of financial means and incentives, and a strong environmental enforcement agency. The grass-roots environmental groups constitute the "green" challenge.

The Green parties in particular emphasize values that promote resilience over stability, response rather than conservation (Cifric 1988 and 1989). Among these, the Greens of Slovenia have given the most comprehensive statement of the Green position on Eastern and Western society. The political objectives of the Slovenian Greens are set forth in its electoral program (Voleni program . . . 1990): democracy, private property, economic competition, and a moderate nationalism. Behind these values lie the postmaterialistic values derived from the West European social movements of the sixties and seventies, especially the German Greens (Milardovic 1989, pp. 70–76; Vukasin 1987, pp. 49–50). As Peter Jamnicar, secretary general of the Party, explained in an interview with the author, the values are global, not related to interests of a specific group as are the values of the old and reformed political parties. In addition, the focus has shifted from quantity to quality of life. For Jamnicar, quality of life is best improved by resolutely opposing all forms of bureaucratization, including technology. In language similar to that of Loebl or Hollings, he sees modern society run by a vast consumption machine. There is little room for spontaneity or creativity. Everything is subsumed under this machine, including the scientific establishment in Slovenia and elsewhere. But society is in the process of trying to escape this inhuman control. The Green alternative calls for a shift to alternative energy sources, to a less hierarchically structured mass society, and to a return to the valuation of the individual and his responsibility for himself and his surroundings. Finally, the Greens call for a shift from seeing progress purely in technological and material terms to seeing progress in psychological, spiritual, and cultural relationships. What modern society needs is space to assert oneself and one's identity in harmony with the environment.

For Jamnicar, the essential break with bureaucratization occurred in Slovenia with the end of the monopoly ideology. This ideology celebrated the destructive aspects of modern society: bureaucracy, technocracy, political hierarchical power. In his view, the West will never adequately resolve its environmental problems until the technocratic monopoly is brought down. But he declined to offer a methodology. In keeping with his party's democratic views, he believes that environmentalists in each country must work out their own strategies.

Jamnicar is convinced that the Greens can offer concrete alternatives to the stalemated organization of modern industrial society. When he and other students made the decision to bring environmental dissent into the open, they determined to reject all aspects of the *totalitarian* bureaucratic structure in favor of new structures and new relationships. The nonhierarchical democratic form of the Green party was one such structure. A new relationship was the penetration of Green sympathizers into all echelons of government. He also stressed the positive approach of the party. "We do not just criticize," he said. "But we try always to offer constructive solutions." He admitted the public might not yet be ready for a postmaterialist ideology, but insisted that the continuation of the old bureaucratic economic system under a new name would ultimately lead to the economic *qua* environmental destruction of Slovenia. Jamnicar believed that "Green" values were the only values capable of mobilizing the public, even though they required the closing of coal mines and nuclear power plants and threatened large-scale unemployment. For him, these phenomena were short-term evils leading to a more harmonious relationship between human society and its natural environment. In his words, "We will have difficulties, but as long as there are men of good will, we will survive."

In sum, the "politicos" see the new elected governments as vehicles through which to obtain the requisite regulations, funding, and technology. The grass-roots organizations remain suspicious of the composition and objectives of any government that practices what Norwegian social scientist Eric Lykke has termed closed or "exclusive" (confined to economically or technologically interested parties) rather than open or "transparent" (democratic) decisionmaking. At the present time, both wings are necessary components to the formation of an aroused, mobilized public strong enough to make its environmental demands carry in the halls of the new governments.

Conclusion

There is abundant evidence that if the East European governments choose to imitate current Western life-styles, they will be locking into environmentally polluting technologies from which it will be exceedingly costly to escape. The wiser course might be to foster working relationships with the environmental groups seeking a way out of the technological labyrinth. Stated otherwise, the governments should forego the industry–government technological union, which has characterized socialism and "capitalism" alike, in favor of building a more broadly based democratic alliance with consumer and environmental groups concerned about the global future of the planet. Western governments are already showing their support of sound environmental problem solving through the funding of pollution-control projects, environmental management training, and legislative education. But it is even more important that they promote democracy. If the new alliance is forged, sustainable development in a clean environment may become a reality in Eastern Europe instead of an unattainable dream. The environmental surprise that helped bring down the socialist monoculture in the East can teach us all a lesson.

References

Bednyi, M. C. 1984. *Demograficheskie faktory zdorov'ia*. Moscow: Finansy i statistika.

Boehmer-Christiansen, Sonjma. 1990. "Putting on the Brakes: Curbing Auto Emissions in Europe." *Environment,* vol. 32, no. 6 (July–August 1990): pp. 16–20, 34–39.

Brooks, H. 1973. "The State of the Art: Technology Assessment as a Process." *Social Sciences Journal,* vol. 25, pp. 247–56.

———. 1986. "The Typology of Surprises in Technology, Institutions, and Development." In *Sustainable Development of the Biosphere,* ed. W. C. Clark and R. E. Munn, pp. 325–50. New York and London: Cambridge University Press.

Budaj, Jan, ed. 1987. *Bratislava nahlas*. Bratislava: Zakladne organizacie slovenskeho zvazu ochrancov prirody a krajiny SZOPK, nos. 6, 13.

Caldwell, Lynton K. 1973. "The Changing Structure of International Policy: Needs and Alternatives." In *Environmental Policy: Concepts and International Implications,* ed. Albert E. Utton and Daniel H. Henning. New York: Praeger.

Chowder, Ken. 1990. "Can We Afford the Wilderness?" *Modern Maturity* (June–July): pp. 61–64.

Cifric, Ivan, ed. 1988. *Drustvo i ekoloska kriza.* Zagreb: Sociolosko drustvo Hrvatske.

————, ed. 1989. *Ekoloske dileme.* Zagreb: Sociolosko drustvo Hrvatske.

"Cleaning Up, A Survey of Industry and the Environment." 1990. *The Economist,* 8 September 1990, pp. 1–26.

Czechoslovak Academy of Sciences. 1989. *Stav a Vyvoj zivnotniho prostredi v Ceskoslovensku.* Ekologicka sekce Ceskoslovenske biologicke spolecnosti pri CSAV a Ekologicka pracovni skupina Rady ekonomickeho vyzkumu CSSR ve spolupraci s Ceskym svazem ochrancu prirody v Ceskych Budejovicich a Praze. Ceske Budejovice and Prague: Cesky svaz ochrancu prirody.

————. 1990. Ekologicka sekce Ceskoslovenske biologicke spolecnosti pri CSAV. Konference "Strategie setrvaleho rozvoje v CSSR," Prague, 31 January 1990–2 February 1990. Collection of papers. Mimeographed.

Czech Republic. 1990. "Koncepce ekologicke politiky Ceske republiky: 'Duhova kniga'," *Uzemni planovani a urbanismus,* vol. 17, no. 3. (Ministerstvo zivotniho prostredi. Pracovni verze pro verejnou diskusi.)

Davis, Ged. R. 1990. "Energy for Planet Earth." *Scientific American,* vol. 263, no. 1, pp. 55–74.

El Serafy, Salah, and Ernst Lutz. 1989. "Environmental and Natural Resource Accounting." In *Environmental Management and Economic Development,* ed. Gunter Schramm and Jeremy J. Warford. Published for the World Bank. Baltimore and London: Johns Hopkins University Press.

EnSol90. 1990. Conference Proceedings. Santa Clara Convention Center, Santa Clara, CA. 12–14 September 1990.

Fri, Robert W. 1990. "Energy and the Environment: Barriers to Action." *Forum,* vol. 5, no.3 (Fall 1990): pp. 5–15.

Galbraith, J. K. 1990. "Why the Right is Wrong." *Manchester Guardian Weekly,* 4 February 1990, p. 10.

Hankiss, Elmer. 1989. "Reforms and the Conversion of Power." *East European Reporter,* vol. 3, no. 4 (Spring–Summer 1989): pp. 8–9.

————. 1990. "In Search of a Paradigm." *Daedalus,* vol. 119, no. 1 (Winter 1990): pp. 183–214.

Hinrichsen, Don, and Gyorgy Enyedi, eds. 1990. *State of the Hungarian Environment.* Budapest: Hungarian Academy of Sciences, Ministry for Environment and Water Management and the Hungarian Central Statistical Office.

Hodge, A. Trevor. 1990. "A Roman Factory." *Scientific American,* vol. 263, no. 5 (November–December 1990): pp 106–11.

Hollings, C. S. 1986. "The Resilience of Terrestrial Ecosystems: Local Surprise and Global Change." In *Sustainable Development of the Biosphere,* ed. W. C. Clark and R. E. Munn, pp. 292–320. New York and London: Cambridge University Press.

Jancar, Barbara. 1987. *Environmental Management in the USSR and Yugoslavia: Structure and Regulation in Communist Federal States.* Durham, NC: Duke University Press.

————. 1989. "Pollution in the Communist World." *World & I* (June 1989): pp. 23–28.

————. 1990a. "United States–East European Environmental Exchange." *International Environmental Affairs,* vol. 2, no. 1 (Winter 1990): pp. 40–66.

————. 1990b. "East European Environment" *Harvard International Review* (Summer): pp. 13–18, 58–60.

————. 1990c. "The East European Environmental Movement and the Transition to Democracy." Paper delivered at the International Slavic Conference, Harrogate, England, July 1990. Mimeographed.

Komarek, Valtr, et al. 1988. *Prognoze a program.* Prague: Academia Praha.

Leonard, H. Jeffrey. 1988. *Pollution and the Struggle for the World Product: Multinational Corporations, Environment, and International Comparative Advantage.* New York and London: Cambridge University Press.

Loebl, Eugen. 1976. *Humanomics: How We Can Make the Economy Serve Us—Not Destroy Us.* New York: Random House.

Lovelock, J. 1979. *Gaia: A New Look at Life on Earth.* New York: Oxford University Press.

Matas, Mate, Viktor Simoncic, and Slavko Sobot. 1989. *Zastita okoline danas za sutra.* Zagreb: Skolska knjiga.

Milardovic, Andjelko. 1989. *Spontanost i institucionalnost.* Belgrade: Kairos.

Mische, Patricia M. 1989. "Ecological Security and the Need to Reconceptualize Sovereignty." *Alternatives,* vol. 14, no. 4, pp. 389–427.

Moldan, Bedrich, ed. 1990. *Zivotni prostredi Ceske republiky: Vyvoj a stav do konce roku 1989.* Czech Republic: Ministerstvo zivotniho prostredi ve spolupraci se s. p. Terplan a s pouzitim materialu Ekologicke sekce Ceskoslovenske biologicke spolecnosti pri CSAV. Prague: Academia Praha.

Orr, David W., and Marvin S. Soroos, eds. 1979. *The Global Predicament.* Chapel Hill: University of North Carolina Press.

Pearce, D. W. 1976. *Environmental Economics.* New York and London: Longman.

————. 1985. "Sustainable Futures: Economics and the Environment." Inaugural lecture, Department of Economics, University College, London, 5 December.

Redclift, Michael. 1987. *Sustainable Development: Exploring the Contradictions.* London and New York: Methuen.

Republic of Croatia. 1990. Republicki komitet za gradevinarstvo, stambeni i komualne poslove i zastitu covjekove okoline and Zavod za prostorno uredenje i zastitu covjekove okoline (1990). *Bilten,* no 1.

Riha, Josef, ed. 1980. *Proceedings of the Ninth Conference on the Biosphere.* Prague: Dum Technicky CSVT.

"Shades of Green, Beyond Earth Day: Ten Views on Where to Go from Here." 1990. *Utne Reader,* no. 40 (July–August 1990): pp. 50–65.

Smith, D. A., and K. Vodden. 1989. "Global Environmental Policy: The Case of Ozone Depletion." *Canadian Public Policy,* vol. 15, no. 4 (December): pp. 423–436.

Socialist Federal Republic of Yugoslavia and Republic of Slovenia. Skupscine. 1990. "Porocilo o stanju okolja v SR Sloveniji," *Porocevalec,* vol. 16, nos. 5/I (20 February 1990), 5/II (20 February 1990), and 7 (8 March 1990).

Timmerman, P. 1986. "Mythology and Surprise in the Sustainable Development of the Bioshpere." In *Sustainable Development of the Biosphere,* ed., W. C. Clark and R. E. Munn, pp. 435–54. New York and London: Cambridge University Press.

Torrens, Ian M. 1990. "Developing Clean Coal Technologies." *Environment,* vol. 32, no. 6 (July–August): pp. 10–15, 28–33.

Vavrousek, J., et al. 1990. *Ekologicke programy a projekty CSFR.* Czechoslovak Federal Republic: Statni komise pro vedeckotechnicky a investicni rozvoj, Ministerstvo zivotniho prostredi Ceske republiky, Slovenska komisia pre zivotne prostredie. 2d Draft, as of July 11, 1990.

Venerable, Grant. 1987. *The Paradox of the Silicon Savior: Charting the Reformation of the High-Tech Super-State.* San Francisco: MVM Productions.

Vintrova, Ruzena, Jan Klacek, and Vaclav Kupka. 1980. "Ekonimicky rust v CSSR, jeho bariery a efektivnost." *Politicka ekonomie,* pp. 29–42.

"Voleni program Zelenih Slovenije." 1990. Mimeographed.

Vukasin, Pavlovic. 1987. *Poredak i alternative.* Niksic: Universitetska rijec, Centar za marksisticko obrazovanje.

Watson, J. Wreford, and Timothy O'Riordan, eds. 1976. *The American Environment: Perceptions & Policies.* London and New York: John Wiley & Sons.

Index

Milton Keynes UK
Ingram Content Group UK Ltd.
UKHW031134141024
449569UK00006B/184